KU-647-684

Contents

Preface

This volume is the second major book in an ongoing series of publications, articles, conferences, and seminars on the topic of the Economic Impact of Worksite Health Promotion, sponsored by the Association for Worksite Health Promotion (AWHP) as a service to the profession. At the Board of Directors meeting in Stamford, Connecticut in December, 1986, this writer, as the newly appointed Director of Education for the organization, secured support from the Board to establish this topic as the subject of an ongoing investigation on behalf of the membership. With increasing frequency, health promotion providers were being challenged by company managers to provide evidence of the economic as well as noneconomic effectiveness of these programs. The "bottom line" question became one of importance to all practitioners.

The first attempt at providing a response to this query was a review of the literature published by this writer in the association journal, *Fitness in Business*, in October, 1987. This effort clarified the magnitude of the problem: of the estimated 12,000 companies with health promotion activities under way at that time, only 10 had reported their outcomes in the peer-reviewed literature. Consequently, in order to stimulate the implementation of such program evaluations, a consensus symposium, jointly sponsored by AWHP and Parke/Davis Corporation, was convened in April, 1990, at the Texas College of Osteopathic Medicine in Fort Worth, bringing together the leading scholars in the field to address this issue. The proceedings of this meeting were collected and published by AWHP as a "white paper" in August, 1990. The annual conference of the association that year established a continuing Economic Impact Track, sponsored by American Corporate Health Programs, Inc., in which the most recent findings in the area were presented.

In January, 1992, Joseph Opatz was appointed vice president for Special Projects of AWHP, and he accepted the challenge of editing the first anthology in this series, which was published in August, 1993. In the meantime, a second Consensus Symposium, sponsored by Blue Cross/Blue Shield of Western New York and the Wellness Institute of Buffalo, was held in Buffalo in April, 1993. The proceedings of that conference and the Consensus Statement on the Economic Impact of Worksite Health Promotion are contained herein.

At the Buffalo conference, the faculty and several other industry leaders in attendance created an ad hoc committee to develop a position paper regarding worksite health promotion in health care reform. Janet Edmunson, then President of the Association, joined with several other industry leaders in formally presenting the resultant paper to the Task Force on Health Care Reform in May, 1993. After several revisions, that paper continues to focus attention on the demonstrated economic benefit of worksite health promotion, and urges the policy makers in this country to ensure that this movement continues to receive support. (The paper is reproduced in its October, 1993, form in the appendix of this volume.) In addition, the authors of that paper formalized their efforts and expanded their breadth to include other individuals and associations in the profession and formed the Worksite Health Promotion Alliance, which continues to cite both the economic and noneconomic benefits of worksite health promotion as a basis for its inclusion in the reformed health care system.

Worksite health promotion does provide such benefits to those who support its implementation and to those who participate in its programs. And the evidence is mounting at an increasing rate. Recently, in a comprehensive survey of the literature, Ken Pelletier pointed out that after collecting 24 total articles on program evaluation in the entire decade of the 1980s, a like number had already appeared in the years 1990 through 1993. AWHP will continue to encourage and foster these studies through its support of publications such as this, conferences dedicated to the topic, and continued opportunities to present the most recent findings at its annual conference.

This author wishes to acknowledge all of the scholars and practitioners whose continuing efforts to study the economic impact of worksite health promotion provide a valuable service to the industry: Joe Opatz, Dave Chenoweth, Bob Karch, Dave Anderson, Dick Huset, Jim Terborg, Ken Warner, Larry Gettman, Roy Shephard, Bill Baun, Steve Blair, Brenda Mitchell, Wendy Lynch, Bill Whitmer, Tom Golaszewski, and D.W. Edington. Special thanks go to Gene Babon, Mary Lee Campbell-Wisley, and Dick Robson for their companies' financial support in this continuing effort, to Phil Haberstro and his staff at the Wellness Institute of Buffalo, New York, and to Susan Presley for her unfailing patience and clerical help in seeing this effort through.

Robert L. Kaman, PhD, FAWHP
November, 1993
Fort Worth, Texas

Chapter 1

The Consensus Statement

An Introduction to the Consensus Statement

Robert L. Kaman

A major focus of the Association for Worksite Health Promotion's second national conference on the Economic Impact of Worksite Health Promotion was the production of the following consensus statement. This document was achieved through the considerable effort of its authors during a closed preconference symposium, April 29-30, 1993, in Buffalo, New York. Each of the authors of this document, top experts in the field of worksite health promotion, submitted a paper for debate and consensus agreement on nine subject areas listed below. Sometimes painful, always intense, and ultimately thorough, this process led to a document that uniquely represents the state of progress in this field to the reader and which provides direction for future studies for the investigator.

Following the consensus symposium, on May 1-2, 1993, the faculty then presented a paper, in open conference, that detailed the current knowledge of the consensus topics:

- Noneconomic benefits of health promotion
- The impact of health promotion on health care utilization
- The impact of health promotion on health care costs
- The impact of health promotion on health behaviors and risks appraisal
- Worksite health promotion and injury
- The impact of worksite health promotion on absenteeism
- Worksite health promotion and productivity
- Health benefits design, and
- Computer simulation of worksite health promotion evaluation.

These papers (printed as chapters 3 to 11 of this volume) form the background for the individual sections of this statement, which the authors agree represents a concise distillation of the current state of knowledge concerning the economic impact of worksite health promotion.

Noneconomic Benefits of Health Promotion

Steven N. Blair

Scientific evidence to support the benefit of health promotion has begun to accumulate only recently. There were a few studies in the scientific literature on these topics in the early part of the present century, but extensive efforts to investigate this issue are concentrated in the past 40 years. The accumulation of well-designed studies is substantial, and we can now make some judgments about the relation between health promotion and health status with a reasonable degree of certainty.

A well-balanced diet, where energy intake is balanced with energy expenditure, is an important contributor to health. Some of the issues are complex and consensus has not been attained for several major questions. It is generally agreed that a diet high in fruit, vegetables, and whole grains, high in fiber, and low in fat (especially saturated fat) and cholesterol is preferred over the typical American diet. A high-fat diet leads to higher risks for atherosclerotic disease, colon cancer, and prostate cancer. There is evidence that high-fat diets increase the risk of hypertension and breast cancer, but consensus has not been achieved. High-fat diets may promote obesity irrespective of total energy intake, but these data are not conclusive. Excessive alcohol intake causes cirrhosis.

Dietary composition probably does not increase the risk of non-insulin-dependent diabetes mellitus, although diet intervention is important in treatment. High-sodium diets may increase the risk of hypertension in susceptible individuals.

Sedentary living habits increase the risk of all-cause mortality, primarily due to higher rates of cardiovascular disease, some cancers, and non-insulin-dependent diabetes mellitus. Low levels of activity probably lead to weight gain. It is established that physical activity increases physical fitness and functional capacity. This health benefit is especially important for elderly individuals, where higher levels of fitness may delay or prevent frailness and relative disability.

There is no doubt that tobacco use causes health problems and that many of the major chronic diseases (cardiovascular disease, many different cancers, and chronic obstructive pulmonary disease) are caused by smoking and use of oral tobacco products. Thousands of studies from epidemiology, pathology, and experimental trials provide an irrefutable body of evidence on the deleterious effects of smoking and other forms of tobacco use, including exposure to environmental tobacco smoke.

There is limited evidence regarding the potential synergism or interaction among various health behaviors. For example, both sedentary lifestyle and a high-fat diet appear to increase the risk of colon cancer. Furthermore, there may be potentiation from a combination of these risk factors, with especially high rates of colon cancer in sedentary individuals who have a high-fat diet.

Data from randomized clinical trials are not available to demonstrate the efficacy of health promotion in preventing disease, and such studies are not likely to be forthcoming due to cost and logistical complexity. Fortunately, recent reports from carefully done prospective observational studies show the potential benefits of health behavior change. These studies provide the strongest evidence now available to support the hypothesis that health promotion reduces disease risk. Individuals who convert from a sedentary to an active lifestyle, stop smoking, maintain weight, and/or avoid becoming hypertensive have a significantly lower risk of cardiovascular disease and all-cause mortality than individuals who do not make these changes.

Summary

Evidence strongly supports the role of lifestyle in determining the risk of developing many major chronic diseases and conditions. Health promotion programs can enhance the adoption of a healthful lifestyle for many participants.

Recommendations for Future Research

It is unlikely that large randomized clinical trials will be conducted to evaluate the efficacy of health promotion on reducing risk of morbidity and mortality. Nevertheless, additional studies are needed to fill existing gaps in knowledge regarding health promotion and disease prevention.

Questions that should be addressed include the following:

1. How are specific dietary components related to disease risk? For example, are vitamins A, E, and C protective for cardiovascular disease and cancer? Does dietary fat play a role in the development of breast cancer, hypertension, and non-insulin-dependent diabetes mellitus?
2. What are the determinants of the age-related weight gain typically seen in populations? What are the independent and interactive effects of diet and physical activity habits relative to weight control?
3. What are the specific dose–response characteristics of the relation of physical activity to disease prevention? Is the intensity of prescribed exercise important, and if so, what is the minimum beneficial intensity? Is there an optimal level of physical activity for health promotion? Are there excessive levels of exercise that increase morbidity and mortality risk?
4. What are the effects of health promotion in specific population groups such as youth, the elderly, women, minorities, and individuals with various health conditions?

5. How do health habits interact in terms of disease risk? For example, can a healthful diet ameliorate some of the harmful effects of stress? Do sedentary living habits accentuate the risks of a poor diet?

The Impact of Health Promotion on Health Care Utilization

Wendy D. Lynch

The impact of health promotion on the utilization of health care services has not been well documented. Although scientists have established a relationship between risks and morbidity, this connection is not clearly reflected in overall health costs or utilization behavior. In part, this is due to the choice and specificity of outcome measures, the complexity of the personal choice to seek health care, and the availability of health care services. The most conclusive studies regarding the influences of health promotion on health care utilization involve interventions aimed directly at the individual's health care decision.

Health behaviors and lifestyle have a direct impact on one's likelihood of developing disease. The causal connection between known risk factors (e.g., smoking, obesity, lack of physical activity, uncontrolled hypertension, lack of safety belt use) and morbidity and mortality rates is well established. There is evidence that the adoption of a healthy lifestyle leads to the compression of morbidity. Clearly, health-enhancing behaviors can extend the period of independent living and reduce the likelihood and delay the onset of many serious and expensive diseases. Consequently, improving health behaviors should reduce the need for medical care.

Most of the literature on health promotion reports economic effects in the form of overall health costs, where the reader must assume that lower expenses implies fewer services. Fewer studies report actual utilization rates, patterns of utilization, or lifestyle-specific utilization.

Studies of risk factors and physician visits have produced mixed results. Risk factors alone seem to account for only a fraction of the variation in utilization behavior. Some evidence regarding the relationship between risk and utilization has come from studies on how high-risk behaviors increase an individual's likelihood of becoming a high-cost health care user. These studies have demonstrated that some risk factors, (e.g., smoking and alcohol consumption) increase one's chances of needing expensive care.

Studies regarding overall utilization of health services reveal a more complex picture of the relationship between risk status and utilization. In some instances low-risk individuals use more health services, especially low-cost services. Furthermore, awareness of risk may actually lead to greater use of medical care in the short term. Thus, while their overall costs may be lower, low-risk individuals may have a greater tendency to use certain types of services. Some of the "overutilization" may be for appropriate health screening or ongoing treatment, but this has not been documented.

Much evidence supports the notion that use of health services is only partially driven by health risk or health status and must be viewed from psychosocial and health system perspectives as well. Perceptions and beliefs have a tremendous impact on utilization. Patients who perceive themselves in poor health or lack confidence in their own ability to manage disease will use more services than those who feel otherwise. Utilization behavior is also driven by the presence and quality of medical benefits and the inherent incentives and disincentives for using them. In addition, provider choices and treatment decisions can have an impact on utilization and cost, independent of true health status. Under a broad definition of health promotion that extends beyond risk reduction, these determinants of utilization can be modified in conjunction with health risks.

Several types of interventions that may not fall under the traditional heading of health promotion or risk reduction have demonstrated an impact on health care utilization. Use of medical self-care education materials reduces utilization and delays treatment of minor conditions, with no measurable deterioration in health. Training on the self-management of arthritis, for example, has led to long-lasting effects on self-efficacy, pain, and cost and frequency of medical services. This self-management training seems to be transferable to many chronic diseases. These and other approaches have led to a reduction in the demand for medical services.

Summary

Most of the health promotion literature has focused on overall cost issues and has not provided detailed information about the types of services and the frequency of utilization. This limits the ability to decipher actual patterns of utilization that result from risk status or changes in risk status.

If health costs and utilization are to be used as outcome measures, researchers, practitioners and policy makers must become more aware of factors, in addition to risk status, that influence health care utilization. Conclusive evidence about the effect of health promotion on utilization will only result from understanding its impact on the need for care, on an individual's perceptions and beliefs about self-management, on the incentive to seek care, and on health promotion's contribution to treatment decisions.

Recommendations for Future Research

To obtain a better understanding of the relationship between health promotion and health care utilization, the following issues should be addressed:

1. Rather than focusing on costs, can specific measures of the type of services and frequency of utilization provide better outcomes for detecting the impact of health promotion?
2. Is there a quantifiable relationship between reduced morbidity and reduced demand for medical service?
3. What is the role of health promotion in managing the demand for medical care? Specifically, do interventions alter morbidity, the perceived need for care, or both?

Health Care Cost

R. William Whitmer

As the cost of medical care has increased to over 14% of the Gross Domestic Product, interest in strategies for cost containment has increased as well. Worksite health promotion is a strategy of cost containment that is gaining interest among those responsible for paying the costs of health care. However, there is uncertainty about cost savings. In a recent survey, 44% of responding company executives indicated they were not convinced that health promotion programs provided cost savings.

Measuring the impact of health promotion on health care cost is difficult because the cost and utilization of medical services are influenced by many factors: Management routinely changes medical plan design; costs are shifted to employees; employers are joining coalitions that extend capitated or global fees; and managed care is becoming more common. Any of these factors can modify medical costs. The measurement of changes in health care costs is also more difficult because of the nature of the costs. Costs vary widely over time, and a small percentage of employees account for a majority of costs, making it difficult to detect any overall average change. Furthermore, overall costs have a limited ability to reflect changes in risk-related cost categories.

The impact of health promotion on health care cost has been demonstrated in several ways, including (a) studies that correlate the number of risk factors with the amount paid for medical claims, with the assumption that future reduction in risk factors will result in reduced costs; (b) studies that show the impact of health promotion on the reduction of risk factors; and (c) studies that describe the impact of health promotion activities on the reduction of medical costs.

Over the past decade at least 44 studies investigated the relationship between health promotion and health care costs. However, most of these studies did not evaluate the other factors that could reduce costs. Further, because of the constraints of the worksite, there are methodological problems and design deficiencies inherent in all existing studies. In some cases, these studies are not conducted under controlled, comparative conditions, and none are conducted with randomized groups.

To enhance the impact on health care cost, there is increasing interest in targeting those with major multiple risk factors. If health screening is the methodology used to identify those at risk, research must be provided to learn how to increase participation rates in health screening and all other phases of the health promotion program.

Summary

Studies indicate that worksite health promotion may produce health care cost savings. However, there are several other factors that may reduce health care costs and utilization. Research is needed to distinguish between the independent influence of health promotion and the synergistic influence of these other factors on health care costs.

Recommendations for Future Research

Future research on the impact of worksite health promotion on health care costs should address the following questions:

1. Can methodologies be developed that provide a more precise estimate of the impact of health promotion on cost savings?
2. Can protocols be designed that increase participation of high-risk individuals in all phases of the health promotion program?
3. Can long-term studies be done to confirm that as medical costs decline, risk factors are minimized or eliminated?
4. Are cost-savings data important to executives who make decisions to start health promotion programs?

Health Behaviors and Risks Appraisal

D.W. Edington

Assessments of health behaviors and risks have been used for several decades. Widespread application occurred following the release of the *Health Risk Appraisal* (HRA) public domain software by the Centers for Disease Control in 1980 and the Carter Center revision in 1987. The risk prediction equations are derived from large-scale epidemiological and clinical studies, behavioral risk surveys, mortality tables, and other well-established data. The calculations are accurate in predicting death rates in large populations. The reliability of the HRA is high when risk projection is the dependent variable. The effectiveness of increased knowledge of personal risk in the process of promoting behavioral change has been difficult to demonstrate because not all members of a population need to make changes in any specific behavior and not all members of the population are at equal "readiness" for change.

The assessment of health behaviors and risks is commonly used in health promotion to assign individuals risk categories and for program and case management strategies. Other major contributions of behaviors and risks assessment include program promotion, awareness, education, and evaluation.

A recent application of health behaviors and risks assessment technology has been to associate behaviors and risks with the costs of health care. In most cases thus far reported, high-risk behaviors have been associated with high costs for health care.

These high-risk/high-cost and low-risk/low-cost relationships can be misleading because not all high-risk individuals translate into high cost or low-risk individuals into low cost. The analysis is complicated by the skewed nature of the cost distribution: 10% of the population accounts for 60% (multiple years) to 80% (single year) of the costs. Clarification of the association between risks and costs will require a new generation of data analyses.

The risk-cost relationship is the rationale for "risk-rating" health insurance. This strategy is usually based on one or more modifiable risk factors. Typical threshold values used to categorize high risks include cholesterol (240mg/100ml of blood or above); blood pressure (90mm Hg diastolic and 140mm Hg systolic); body weight (120% or 130% of the 1958 Metropolitan Life Tables); safety belt use (74% use or less); vigorous physical activity (less than once per week); use of alcohol (21 drinks per week or more); and use of cigarettes (current smoker). The excess health care costs associated with each high-risk behavior range from $150 to $400 per year (1990 dollars). The excess costs are not additive, but there is evidence that increasing numbers of high-risk classifications are associated with up to $1,300 per year excess costs for an individual with six or more high-risk classifications compared to an individual with zero high risks.

Summary

Most high-risk individuals have high health care costs. Now that the efficacy of health promotion interventions has been demonstrated in reducing risk, a pressing issue related to the use of health behaviors and risks assessment in risk-rating is whether reduced risks link to future lower costs. If they are linked, within what time frame? Until this issue is resolved, health promotion should continue to emphasize the educational and motivational value of the assessment of health behaviors and risks rather than focusing on its cost implications.

Recommendations for Future Research

Future research into the relation between health behaviors, health risks, and health care costs should focus on the following questions:

1. To what degree can excess health care costs be attributed to risks?
2. What are the age- and gender-specific criteria for risks and costs?
3. How can risks and costs assessments be improved?
4. Do "risk-rating" health insurance practices reduce costs and risks?
5. Do changes in health risks and behaviors have other economic implications?

Worksite Health Promotion and Injury

David Chenoweth

Most of the economic impact articles published on worksite health promotion focus on one or more of the following outcome measures: employee health status, absenteeism, productivity, health care utilization, and health care costs. Only a few studies have investigated the relationship between worksite health promotion and occupational and nonoccupational injuries. One reason for the scarcity of studies is the separation of payment and reporting systems for injury versus medical care expenses.

Cumulative trauma disorders (CTDs) are the most common type of injury reported by American workers. Back injury, often classified as a CTD, is the most prevalent and second most costly CTD disorder. In fact, approximately 75% of all American workers will experience a back injury—short-term, in most cases—sometime in their working lives. There are also substantial costs for long-term compensable back injuries.

In response to the growing number of CTD's, employers have expanded injury prevention efforts to include prework stretching routines, flexibility enhancement exercises, awareness campaigns, ergonomic interventions, and general and specific physical fitness programs.

Approximately half of the published articles focusing on the impact of worksite health promotion on low back injury include a benefit-cost analysis (BCA).

Benefits are typically quantified in terms of actual or estimated cost-savings attributed to one or more of the following categories:

- Lost time (injury-related absenteeism)
- Medical care costs associated with injuries
- Workers' compensation costs associated with injuries

Summary

Most studies indicate that the preceding worksite health promotion efforts can reduce the incidence, severity, and associated costs of CTD. Nevertheless, most of the CTD-based studies focus primarily on back injuries and do not address other common occupational ailments such as carpal tunnel syndrome, repetitive motion injury, and mental distress.

Recommendations for Future Research

Virtually all of the studies reviewed have some methodological shortcomings that limit their generalizability to other worksites. Thus, to minimize potential drawbacks, future studies should address the following questions:

1. What risk factors predict workers with high injury rates? Do specific risk factors correspond with specific types of injuries?
2. How do the rate and type of injuries on the job compare with injuries incurred during fitness program participation?
3. How can injury data be used to determine the most appropriate injury prevention interventions?
4. Which health promotion interventions are most effective in preventing injuries? Which are most effective in facilitating a speedy recovery and a return to unrestricted work?
5. Are there differences in methods of preventing injuries between small and large worksites?
6. What is the effect of specific health promotion interventions on workers' compensation–based injuries?

The Impact of Worksite Health Promotion Programs on Absenteeism

William B. Baun

Despite considerable improvements in the quality of health care and socioeconomic conditions in this country, absenteeism has been increasing in the workplace. The cost of absenteeism to U.S. industry has been estimated at 30 billion dollars per year. Typical U.S. absenteeism rates range between 2% and 5% (4 to 10 work days) per year, and in manufacturing industries the rate reaches as high as 15% to 20%. An absenteeism rate of 5% could cost a company as much as 8% of salary budget per year.

Absenteeism is typically defined as the condition in which employees are not at work either due to sickness or injury accepted by the employer or for some personal reason. A lack of standardization of definition and measurement presents a challenge for the interpretation of absenteeism data. There is a wide range of variables known to influence rates of absenteeism that must be controlled in research design or analysis. Major factors are gender, age, occupation, and administration of absenteeism policy. Because female employees often assume the role of care giver within the family, and thus usually stay home with sick children, absenteeism rates for women in most work environments are higher than those for men. Older workers tend to have fewer periods of absenteeism, but their duration of absence is longer. Many studies have shown that occupation and company policy can account for as much as a threefold difference in absenteeism rates.

Many articles have been published on the effects of health promotion programming on absenteeism. The majority of these studies are longitudinal and are evenly mixed between large and midsized industrial and white collar populations. The effect of health promotion on absenteeism has also been studied on bus drivers, police staff, postal workers, teachers, and highway workers. The impact of a variety of health promotion interventions has been studied, including smoking cessation, stress management, hypertension management, health risk assessment, and fitness.

Smokers have higher absenteeism rates than nonsmokers or ex-smokers. Few smoking cessation studies have tracked absence, but it appears that these programs can reduce absenteeism. Employees who report high stress have higher absenteeism, compared to employees reporting low stress. Use of employee assistance programs reduces employees' perception of stress and decreases absenteeism. Several studies have shown that individuals with hypertension have higher rates of absence. Hypertension intervention programs have had mixed results due to the "sick person" labeling problem, which may increase absenteeism. Health risk assessment (HRA) instruments help characterize major behavioral risks associated with higher rates of absence. A few studies have shown that an HRA intervention can reduce absence levels.

The impact of fitness levels and programs on absenteeism has been studied in different environments, occupations, and population sizes. Studies clearly show that individuals with higher fitness levels are absent less than individuals with lower fitness levels, with differences ranging from 0.28 days to 2 or more days per year. Increases in fitness are associated with reduced employee absenteeism. Cardiac rehabilitation and healthy back programs have also been shown to reduce employee absenteeism.

Summary

Absenteeism is an economic variable that has an extensive research base in health promotion, unlike productivity, turnover, and employee recruitment. Although much of the research lacks adequate controls or suffers from poor design, the data available provide information that is helpful for improving programming decisions and obtaining commitment from management for program support.

Recommendations for Future Research

Future research on the relation of worksite health promotion programs and employee absenteeism should focus on the following questions:

1. Absenteeism data are positively skewed as are medical cost data. Can techniques be developed that will provide better analyses on the effect of health promotion interventions on employees across levels of absenteeism?
2. Can valid definitions and measures of employee absenteeism be designed and implemented in a cost-effective way?
3. What variables mediate the causal relation between health promotion programs and absenteeism?
4. What is the long-term effect of health promotion programming on absenteeism?

Worksite Health Promotion and Productivity

Roy J. Shephard

Productivity is a function of the relationship between the quality and the quantity of production relative to the input of capital in the form of personnel, materials, and/or equipment. In broad terms, productivity involves not only performance at the worksite but also a reduction of absenteeism, turnover, and industrial disputes. This discussion is limited to productivity on the job.

Clarification of the possible impact of fitness and health promotion programs upon productivity has been hampered by problems of design, including a poor definition of the intervention. Typically, the investigator has offered a variety of health promotion modules, often with moderate aerobic exercise as a major focus. However, published reports also cover the response to brief fitness and relaxation breaks and initiatives where the primary focus has been overall health promotion without specific provision for worksite exercise.

It is not possible to carry out double-blind controlled studies of fitness and health promotion programs. A few studies have made parallel investigations at matched experimental and control worksites, but most interventions have either lacked controls, or, at best, they have compared responses of program participants with nonparticipants. The validity of these findings has thus been compromised by contamination and Hawthorne-type responses and the use of self-reports or supervisor reports of productivity. These reports may be influenced by the attitudes of management and personnel toward the intervention. Self-reported productivity and actual productivity may each be affected independently by changes in mood-state. Other more objective measures of productivity include changes in task errors, the quantity of production, and the time taken per task relative to standard times.

In theory, the performance of heavy physical work could be limited by an inadequate aerobic power, lack of muscle strength and endurance, or problems of thermoregulation. In practice, immediate productivity seems constrained by human performance in only a very few tasks performed under particularly adverse environmental conditions. This immediately limits the potential for improvement of productivity by aerobic or muscular training or by other health promotion measures that increase available aerobic power (for instance, a decrease of obesity or smoking cessation). Nevertheless, chronic fatigue associated with near-limiting work could lead to poor quality output and less direct adverse effects upon productivity, such as employment disputes and accidents.

In light physical and sedentary work, several studies suggest that an exercise break can reduce the number of errors, probably through a recovery from mental or physical strain, relief of boredom, and/or an increase of alertness rather than through an enhancement of physical working capacity. In tasks demanding mental effort, many self-reports refer to greater self-confidence, relief of stress, and substantially improved productivity following implementation of worksite fitness and health promotion programs. However, it is difficult to dissociate such reports from Hawthorne effects and nonspecific responses to favorable changes of mood-state. Gains of worksite productivity are much less readily demonstrated in terms of supervisor ratings and the timed performance of specific tasks.

Summary

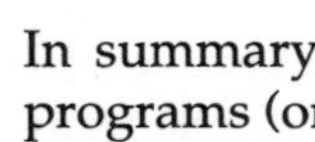

In summary, many companies operating fitness and health promotion programs (or contemplating their introduction) list an increase of productivity as an important anticipated outcome, and some evidence supports such an expectation. Use of better productivity measures is needed to quantify such gains.

Recommendations for Future Research

In researching the possible correlation between employee participation in worksite health promotion programs and productivity, answers should be sought for the following questions:

1. How can worksite indices of individual and corporate productivity be improved, particularly in companies that do not have a tangible and easily measured end-product?
2. Given that an effective dose of exercise cannot be administered in double-blind fashion, what alternative interventions can be developed that have an equal impact on participant attitudes and expectations?
3. What gains of productivity can be shown by objective comparisons of experimental and control worksites if such studies are continued for sufficient time to minimize the immediate Hawthorne impact of the fitness and health promotion intervention?

4. Are any effects of fitness and health promotion programs on productivity related to initial job satisfaction and initial productivity?
5. How do any gains of productivity achieved through fitness and health promotion programs compare with what could be achieved through investment in other initiatives likely to boost production?
6. Which components of health promotion programs make the largest contributions to productivity? Is there a synergism among program components in various occupational settings?

Health Benefits Design

Thomas J. Golaszewski

As the demographic profile and needs of the American family shifted during the latter parts of this century, benefits plans changed. And as plans changed, so did their costs. For example, benefits accounted for approximately 28% of labor costs by 1990, as compared to 17% in 1966. Health care costs specifically represent nearly half of this amount.

In part, these growing health care costs can be traced to the way employee benefits have been designed and administered. Until recently, health care has been largely subsidized by the employer, with few disincentives to deter employees and dependents from accessing and perhaps overutilizing medical care. At the same time, the system offered providers incentives for overutilization. The combination of these factors has had the effect of increasing consumption without enhancing health. In addition, a growing body of research indicates that health care costs are not randomly distributed throughout an employee population. Individual differences in the use of health care exist based on environment, heredity, and lifestyle (those behaviors that the individual presumably chooses to adopt freely). Lifestyle differences also relate to the use of sick leave.

As health care costs skyrocketed during the 1980s, numerous attempts at cost control were implemented. Along with increased contributions from employees, typical programs also adopted limitations on providers, prehospitalization certification, mandatory second opinions, and other administrative controls. Despite these efforts, costs continued to rise, and employers now found health care at the forefront of labor disputes.

Organizations may have contributed to the health care cost crisis by inadvertently subsidizing poor health behaviors with liberal fringe benefits at the expense of employees with good health behavior. This focus has started to change, partly because benefits are no longer so generous, but also because more companies are supporting healthful behavior. However, despite rapid program growth, skepticism persists about health promotion's ability to contain cost. Widespread adoption of healthy behaviors by employees and their dependents remains elusive.

Those implementing programs now believe that the culture of the organization must change if significant numbers of employees are to adopt healthful behaviors. Incentives increasingly are seen as a mechanism to shape organizational cultures and influence behavior. Evidence supporting the effectiveness of incentives in worksite health promotion programs exists. Benefits-related incentives have recently been added to this strategy. For example, some companies provide discounts on health insurance copayments or deductibles if employees meet personal risk-factor guidelines. Others provide employees with rebates if the company's health care costs fall below targeted norms.

Summary

A strong argument can be made for the use of benefits as incentives, although empirical evidence is lacking on their effectiveness. With some reservation, particularly with the use of "risk-rating" of benefits, a growing number of organizations are choosing to incorporate benefit plan changes.

Recommendations for Future Research

New approaches linking health promotion to benefit designs lead to questions of their efficacy and impact. These questions include the following:

1. Do benefit incentives alter the use of health promotion services? With or without an increase in services, do they improve health behavior, decrease morbidity, and affect utilization and health care costs, absenteeism, and other economic variables?
2. What are the effects of benefit incentives on employee morale and satisfaction with the benefits plan?
3. Do benefit incentives affect employee recruitment and/or retention? Do they affect the recruitment of health-oriented employees?
4. What are the legal and ethical implications to varying benefits distribution based on individual health criteria?

Computer Simulation

James R. Terborg

Research has documented the beneficial effects of worksite health promotion programs on employee behavior and health risk status. Data further suggest that such programs can be cost-effective compared to nonworksite programs. Less is known about economic benefits that may result from presumed reductions in health care costs, absenteeism, accidents, and turnover and improvements in employee morale and productivity. Although, statistically, reliable associations have been reported between employee health behavior and organizational outcomes, considerable uncertainty exists regarding the cost-benefits of worksite health promotion programs.

The primary reasons for this uncertainty are the paucity of rigorous experimental designs and the difficulty in data acquisition. The importance of experimentation is widely recognized, but advances in knowledge are limited by constraints placed on researchers operating in the worksite. Even if long-term access to a sample of worksites were obtained and worksites were randomly assigned to treatment conditions, the costs of evaluation would be substantial. Companies wishing to make a specific evaluation of their worksite program might find that the cost of such analyses would exceed the presumed savings. Furthermore, even if the results of such studies were available, the results might not generalize to other worksites with different employee demographics, medical plans, and turnover rates. One promising approach to this problem is the development and use of computer simulations.

A computer simulation is a model of a system or process. By conducting experiments with the model, it is possible to begin to understand how the system functions and to evaluate various strategies for its operation. Computer simulations have several components: the assumptions that underlie the model; the independent, dependent, and control variables; and the algorithms that describe the relationships among these variables.

Computer simulations are not substitutes for empirical research. Rather, simulations complement such research when empirical studies are costly, difficult, unusually complex, or unethical.

Simulations are common in the physical and biological sciences and in operations research, but they have not been widely used in the social and behavioral sciences. Given the current knowledge base regarding worksite health promotion programs and the difficulty in conducting more comprehensive empirical research, the design and use of computer simulations are highly appropriate. Some knowledge already exists on the costs and benefits associated with health promotion programs. These data can be used as starting points. By systematically varying the number and magnitude of independent and control variables, it is possible to project various outcomes and the probability of each under different scenarios.

Computer simulations are highly efficient for asking "What if?" questions. Sensitivity analysis examines the impact of changes in one variable at a time, for example, the impact of different smoking cessation rates on projected long-term reductions in health care costs. Scenario analysis examines the dependent variable under different sets of independent variable values, for example, best case, worst case, and base case scenarios. Break-even analysis focuses on the values certain variables must take to satisfy constraints placed on the model; for example, given specified fixed and variable costs and fixed and variable benefits, how many employees must participate in a program to achieve a cost-benefit ratio of 1.00? Finally, capital budgeting techniques such as net present value analysis can easily be incorporated. All of these features facilitate experimentation, which contributes to theory building and policy planning.

Two computer simulations have been done on the economic consequences of health promotion programs at the worksite, one using hypothetical data and the other using data from a specific company. Both reported projected benefits exceeding program costs. The results are consistent with cost-benefit ratios estimated by simpler approaches as reported in published empirical studies. The simulations suggest that 75% or more of the total economic benefit of health promotion at the worksite can be attributed to changes in productivity, absenteeism, and turnover and that 25% or less of the total benefit stems from reductions in medical costs.

Results can vary substantially, however, depending on assumptions regarding the dollar value of changes in productivity. Furthermore, sensitivity analyses show that small changes in productivity, absenteeism, and turnover have large effects on projected economic benefits. In contrast, sensitivity analyses suggest that changes in medical cost inflation, the discount rate, and retiree health cost have relatively little impact on total economic benefits. Other variables that have a moderate to high impact on total economic benefits include average employee age, employee health status, participation rates in the program, and program effectiveness. The computer simulations that have been done confirm that characteristics of the company, the health promotion program, and employees determine whether or not the dollar value of benefits exceeds the program costs. Consequently, results may not generalize from one company to another.

Summary

Computer simulations that attempt to model worksite health promotion programs suggest that such programs may be cost beneficial, that economic benefits primarily come from increases in productivity and less so from decreases in health care costs, and that program effects vary across different organizations.

Recommendations for Future Research

In order to make computer simulation a truer picture of the costs and benefits of worksite health promotion programs, the following questions need to be addressed:

1. What are the assumptions that need to be made prior to the design of a computer simulation?

2. What are the most important variables that need to be considered as inputs, control variables, and outputs?
3. What algorithms or rules should be specified for linking input and control variables to outputs?
4. What range of values do we assign to input and control variables? How do we know these estimates are valid?
5. Can we validate the simulation by demonstrating correspondence between outcomes empirically derived and outcomes predicted by the model?
6. Are results of computer simulations useful as decision aids to executives and policy planners who make decisions regarding health promotion programs at the worksite?
7. Do computer simulations help researchers identify important research questions that are amenable to subsequent empirical verification?
8. Do computer simulations provide for a realistic understanding of the structure and internal processes of health promotion programs at the worksite?

Chapter 2

Introduction to the Chapters in Support of the Consensus Statement

Robert L. Kaman

Worksite health promotion programs, which offer a variety of activities designed to enhance the health and fitness of their participants, originated as an expression of paternalistic support by company owners for their employees. As these programs evolved into sophisticated strategies of primary prevention and required increasingly large budgets to sustain, managers began to question their fiscal value to the company. Was the substantial investment in facilities, staff, and operations justified by more than just the anticipated gratitude of the workforce? If participants did become healthier and more productive, did these changes translate into reduced costs for personnel or into net gains in profitability? Answers to these questions began to appear in the literature over 20 years ago and have subsequently evolved into a rigorously defined framework of social science research. These studies generally support the notion that worksite health promotion programs are both cost-beneficial and cost-effective for the company as well as for the employee as a strategy for improving health and reducing costs. Despite these promising results, conclusive evidence ensuring the future of worksite health promotion remains elusive, and program outcome evaluation studies continue.

An immediate threat to the future of worksite health promotion may lie in the changes in the American system of health care that could occur as a result of the recommendations of the Task Force on Health Care Reform. A primary goal of the reform, as frequently stated by politicians and experts alike, is to extend access to medical care to every American. This so-called "universal" access will be paid for by some form of additional taxation, and the lion's share of this cost may fall on the shoulders of the American entrepreneur. Plans call for increased health insurance rates, or a new payroll tax, imposed on the employer and the employee alike. If the payments are based on a community rating system instead of on an experience rating system, all economic incentive for introducing health promotion programs, or for sustaining established programs, will be lost.

During the past 25 years, many American companies have independently created a system of health promotion activities at the worksite, without government mandate or incentive, as a strategy for improving the health and productivity of their workers and to reduce health care costs. American businesses have become convinced that company-sponsored health promotion and fitness programs can be an effective cost-containment strategy. Evidence continues to mount that health promotion program participants take less sick leave and use fewer health care dollars, stay with the company longer, and are more productive than nonparticipants. Companies realize actual financial benefits, in part, by health insurance rates that stay flat or which rise at slower rates than those of companies without such programs. Health insurance financing based on a community rating system that would even out insurance rates for all companies eliminates the incentive for companies to invest in programs that will improve worker health. Experience rating, on the other hand, allows a company to set rates based on its own health care cost history, thereby establishing a rationale for investing in employee health promotion programs. Clearly, if a national health care reform program includes community rating for health insurance, worksite health promotion programs become vulnerable. In the meantime, economic outcome analyses of worksite health promotion programs continue.

The economic impact of worksite health promotion programs is measured by changes in worker health, health care utilization, health care costs, health risks, injury rates, absenteeism, and productivity. In addition, companies have explored the extent of the impact of changing health benefits plans to reward healthy behaviors. Investigators have also established computer simulations of the economic impact of health promotion programs, using normative data instead of actual measurement, to project savings from such programs. Despite a purported "academic conservatism" (Fries et al., 1993), the overwhelming body of evidence from these studies continues to support the notion that worksite health promotion programs are indeed associated with improvements in each of the areas measured.

What follows briefly introduces each of those areas of measurement, as discussed in greater detail in the ensuing chapters, in support of the validity of a positive economic benefit for companies that institute such programs. Each chapter is discussed in its order of appearance in the book and is headed by a summary of the consensus statement for that topic.

Noneconomic Health Benefits of Worksite Health Promotion Programs

Health promotion programs are effective in helping employees adopt more healthful lifestyles marked by increased physical activity, avoidance of tobacco, and a diet high in fruits, vegetables, and whole grains and low in fat.

If there is a link between lifestyle factors and chronic disease, then changes in lifestyle should have measurable impact on the incidence and severity of those diseases. Indeed, chronic diseases are the major cause of death in the United States today, and lifestyle factors have been linked with their occurrence. Changes in lifestyle, supported by participation in health promotion programs, should be followed by changes in the incidence of cardiovascular disease, stroke, cancer, and other chronic diseases. The health promotion intervention model follows the logic that when people change their behaviors from those that facilitate the onset of illness to those that promote health, an improvement in health should ensue.

Four major health habits linked to disease may be examined: diet, physical activity, smoking, and weight control. Improved diet should lead to reductions in the incidence of cancer, cardiovascular disease, cirrhosis, non-insulin-dependent diabetes mellitus, and obesity. Increased physical activity should result in reductions in cardiovascular disease, cancer, non-insulin-dependent diabetes mellitus, and obesity and in improvements in functional capacity. Weight control similarly should result in reductions in cardiovascular disease, stroke, non-insulin-dependent diabetes mellitus, hypertension, and arthritis. Elimination of cigarette smoking should result in reductions in most of the diseases just listed, but its major impact is seen in a decrease in cardiovascular disease and lung cancer.

In chapter 3, Blair cites three prospective observational studies available to establish the link between improved health behaviors and improved health. In the Alameda County (CA) Study, it was shown that individuals 50 years of age and older who quit smoking or who increased their physical activity had lower risk for all-cause mortality than those who did not. In studies conducted by surveys of alumni from the University of Pennsylvania and Harvard College, the evidence again supports the notion that risk for heart disease and cancer is reduced by as much as 50% in those individuals who quit smoking or increased their physical activity.

Clearly, these studies, and others similar to them, reinforce the intuitive logic that healthy behaviors improve health and that these healthy behaviors are causally related to such changes. It is not a major deductive leap to reason, therefore, that noneconomic improvements in health, supported by participation in health promotion programs, may engender economic benefits as well—if these improvements in health result in lower costs for and utilization of health care benefits and improvements in worker functional capacity.

The Impact of Health Promotion on Health Care Utilization

Utilization of health care is a complex behavior that is only partly related to morbidity. Thus, risk reduction has limited potential to impact the decision to seek care. Health promotion has a greater potential for reducing utilization through interventions that promote appropriate self-care and self-management of chronic disease.

The usual outcome measure of the impact of health promotion programs, and its anticipated improvements in health, is a reduction in health care costs. Indeed, there is much evidence that suggests that this inverse relationship is real and that improved health results in reduced costs for health care. Recently, observations on health care utilization rates in otherwise healthy populations suggest that the relationship between health behavior and health care cost is not as straightforward as presumed. Morbidity represents only one of many factors affecting utilization rates; therefore, the presumption that improved health results in reduced utilization and costs may be erroneous.

In chapter 4, Lynch describes the utilization of health care services in the traditional economic context of supply and demand. Hence, if this model were correct, one would expect restricted access or lowered health risks (or morbidity) to reduce utilization. Early studies, however, suggested that need, or morbidity as measured in terms of symptoms, could not account for the variability in utilization observed. Instead, the demand component for utilization contains at least three elements in addition to morbidity: perceived need, patient preferences, and moral hazard. (The supply side is just as complex but will not be discussed here.) Perceived need includes a variety of personal, cultural, and social influences. Patient preference includes issues of choice and cost considerations. Moral hazard refers to some external influence on behavior, such as incentives, or simply an excuse to leave work. Therefore, since health promotion addresses only the need for services (morbidity), it becomes self-limiting as a factor in reducing health care costs. If actual morbidity accounts for only 12% of the variability of utilization, then health promotion programs that are highly successful in improving health will yield only modest reductions in health care costs.

Health promotion therefore should expand its reach to specifically address a reduction in utilization, targeting utilization behaviors as well as health risks. Proper use of the medical care system, effective strategies in self-care, and improving one's self-efficacy (confidence in one's own ability to manage health) are all compatible with worksite health promotion. In the meantime, measurements of health promotion programs' impact should not ignore the complexities of health care utilization.

The Impact of Health Promotion on Health Care Costs

Studies indicate that worksite health promotion may produce health care cost savings. However, there are several other factors that may reduce health care costs and utilization. Research is needed to distinguish between the independent influence of health promotion and the synergistic influence of these other factors on health care costs.

Perhaps the greatest challenge facing effective worksite health promotion program evaluation is the nonstandardized nature of its various services. The variability extends from actual services offered to the differences within program formats to the size of the exercise facilities when provided on company premises. No two programs are alike, nor are the employee groups and the dependents they serve. This characteristic dilemma facing social science research in general is magnified in health promotion evaluation efforts since there exists an urgency to demonstrate that these programs do in fact save money. We may wonder why such an expectation exists for this health-related preventive service and not for other health care practices. The expectation for surgical procedures is that they be safe; the expectation for medical procedures is that they be safe and effective; the expectation for health promotion services is that they be safe, effective, *and* cost effective (Warner, 1992).

Nevertheless, no matter how the expectation for cost-effectiveness for health promotion arose, it exists in fact, and studies continue to appear in the literature documenting some aspect of the economic impact of these programs. In all of these studies, outcomes are consistently positive (that is, there is an excess of benefit compared to program cost, when measured as cost-effectiveness or cost-benefit) or, at the least, they break even. In chapter 5, Whitmer presents several such studies to support this view, focusing on three aspects of health promotion services: medical screening, the relationship of high-risk behaviors to high medical costs, and cost-benefit analyses of several different programs.

Generally, medical screening, when offered at the workplace as a part of a health promotion program, is conducted on a voluntary basis, with characteristically low (15% to 28%) participation rates as a result. Whitmer cites two different strategies that show surprising success in increasing participation rates to as high as 90% to 95% of eligible employees. By simply scheduling all employees for the voluntary screening, whether they signed up or not, and assigning them an appointment, participation increased to 90%; by tying participation to eligibility for health insurance, participation rose to 95%. In the latter case, fully 13% of those tested had heretofore undetected medical problems requiring attention for early treatment. Although cost savings were not calculated in this case, it is clear that the prevention of one heart attack or by-pass surgery by early detection could save thousands of dollars.

Increasing the participation rates of medical screening programs eliminates the self-selection bias observed in voluntary programs. Workers with unhealthy behaviors tend to avoid such services, and these high-risk individuals generally become high-cost users of medical care services.

Whitmer, in chapter 5, and Edington, in chapter 6, detail several studies in which this relationship between risk and cost is evaluated. Once again, the overwhelming body of evidence supports the notion that high-risk behaviors result in higher health care costs.

Finally, program effectiveness, measured as benefit-to-cost in dollar amounts, was tabulated from eight widely varied health promotion programs. Although the external validity of this type of data is limited, outcomes ranged from break even, to a positive benefit-to-cost ratio of $9.33. The average was a return of $3.57 for every dollar invested, measured only as a reduction of health care costs for participants compared to nonparticipating employees. Recently, Pelletier (Pelletier, 1993) reviewed 47 worksite health promotion programs, and while not all of them conducted benefit-to-cost evaluations, those that did showed a positive return on investment, and all reported some positive economic, physiologic, or sociologic outcome. Nevertheless, the causal link between participation in health promotion programs and reduced cost for the individual or the company remains an elusive goal. That conclusion awaits the development of standardization in programming and more rigorous control of experimental variables.

The Impact of Health Promotion on Health Behaviors and Risks

The profession of health promotion has demonstrated its efficacy in reducing the risks associated with the known precursors to disease. Most studies now agree that high-risk individuals are high cost. The current challenge to the profession is to link changes in high-risk behaviors to cost outcomes.

The evidence for the direct relationship between high-risk behaviors and high medical costs, and low-risk behaviors and low medical costs is most compelling. Notwithstanding the nondefinitive quantification of specific risk behaviors and the etiology of specific diseases, most experts agree that approximately half of all illness is risk-related, and therefore preventable. By extension, some amount of health care costs, approaching 50%, should be amenable to reduction by eliminating high-risk behaviors. The greatest impetus toward fostering an awareness of this relationship was the development of the Center for Disease Control's Health Risk Appraisal in the early 1970s. This instrument, and others that followed, enabled a simple assessment of identifiable risk factors to be conducted, and subsequently, the relationship between the multiplicity and severity of those risk factors and morbidity and mortality could be defined.

Edington, in chapter 6, cites the principal examples of those relationships. In general, according to several studies, the greater number of risk factors present in the lifestyle of an individual, the greater the morbidity, mortality, and cost for medical services that occur. When measured at the worksite, as part of health promotion program risk assessment efforts, strong evidence exists that smoking, obesity, hypertension, lack of exercise, high blood cholesterol, and the excessive use of alcohol and other drugs are independently associated with increased health care and related costs for the company.

The resultant question, then, is whether individuals who change their behaviors to more healthy lifestyles will also realize measurable improvements in health and decrease utilization and incurred health care costs. Edington reviews the evidence that addresses that question. Participants in weight loss programs, smoking cessation, hypertension treatment, cholesterol screening, drug screening, and physical fitness programs in fact show lower costs than nonparticipants, although many of the studies do not display the scientific rigor necessary to establish causality rather than inference. Nevertheless, evidence has not been forthcoming to disprove these observations. Health promotion programs should therefore continue to emphasize the educational, motivational, and health value of lowered risk, while continuing to evaluate the postulated relationship between lowered health risk and lowered cost.

Worksite Health Promotion Programs and Injury

Specific types of worksite health promotion interventions such as prework stretching routines, flexibility enhancement exercises, awareness campaigns, and physical fitness programs can reduce the incidence, severity, and associated costs of injuries, especially low back injury.

The impact of worksite health promotion programs on the incidence and outcomes of work-related injuries has not received the attention that has been afforded to noninjury-related health and fitness. Generally, since such medical problems are administered through a different system (workers' compensation rather than health insurance), the possible relationship between health promotion, fitness, and injuries has not been explored. Recent events have caused investigators to revisit this issue, and the initial results are not surprising.

First, workers' compensation costs have, like other health care costs borne by the employer, escalated to alarming levels. Second, there has been an apparent trend away from traumatic injuries (perhaps as a result of OSHA-mandated safety regulations) and toward the chronic, so-called cumulative trauma disorders (CTDs) such as low back pain and carpal tunnel syndrome. Third, early studies have suggested that there is an inverse relationship between both aerobic and musculoskeletal fitness and the incidence and severity of injuries. These events have led to an industry-wide inquiry into the role that worksite health promotion could play in addressing this problem.

Chenoweth, in chapter 7, surveys the recent developments in this field, and his observations support the view that worksite health promotion can indeed play a significant role in reducing the incidence and severity of on-the-job injuries. In one study, the improvement in fitness among participants in a worksite program was correlated with a 45.7% decrease in major medical costs, and a 31.7% drop in disability costs compared to those costs for the work force as a whole. In another study, after 6 months in a company-sponsored program, one participant group's (back) fitness levels rose 14.2%, while injury-related absences dropped .25 days per year. Another group of participants in the same study showed a decrease in lost-time accidents of 44%, compared to nonparticipants. Several studies have shown that fire fighters who improve their fitness levels have significantly lower rates of job-related injuries.

Perhaps of more significance to the worker and employer alike is that even modest exercise programs are effective in reducing job-related injuries and their associated costs. Brief (5-minute) stretching routines were sufficient to eliminate newly diagnosed CTDs from a work force of data-entry operators and to improve overall productivity by 25%. In another study, weekly back-care education and exercise sessions virtually eliminated back injuries among staff nurses in an eastern hospital. Other studies have suggested that more comprehensive programs, addressing the complex etiology of CTDs, may be more effective in reducing their occurrence.

Even as the evidence mounts in support of this approach to on-the-job injury prevention, investigators have recognized the limited external validity that these studies, like the studies of health promotion's impact on health, may have. Despite that cautionary note, worksite health promotion programs specifically targeting this newly discovered opportunity for effectiveness will be expanded as more employers recognize their utility in addressing this problem.

The Impact of Worksite Health Promotion on Absenteeism

Absenteeism is an economic variable that has an extensive research base in health promotion, and although much of the work lacks the rigors of good research, the data available provide information that is helpful for improving programming decisions and obtaining commitment from management.

Absenteeism as a parameter of evaluation of worksite health promotion programs is a logical but often misleading indicator. If employees who participate in health promotion programs become healthier, they may be absent less frequently. They should also use fewer health care resources and expend lower health care costs. By being absent less, they should be more productive. However, just as health care utilization is complex, so too is absenteeism. Indeed, absence rates in many companies, no matter how healthy the work force, are relatively low (2% to 7%). Changes may be difficult to detect using often imprecise methods (self-reporting, personnel records, etc.). Therefore, despite an "effective" health promotion program, expected decreases in absenteeism rates may not be observed.

Baun, in chapter 8, describes the complexities of absenteeism measurement as a human resource variable and cites several studies that have attempted to use this parameter as a measure of health promotion effectiveness. There are three major categories of factors that influence workplace absences: personal, attitudinal, and organizational. Thus, variables such as gender and race (personal), job satisfaction (attitudinal), and work unit size (organizational) all affect rates of absence independent of health status or participation in health promotion programs. Controlling for this "background noise" is the first challenge facing the investigator.

Even when such effects are accounted for, the actual measurement of absence data is the next hurdle to be overcome. Of the 41 different absence measures appearing in the literature, most studies describe absence as the *incidence rate* (number of workers absent/total employees × 100). The data itself is difficult to acquire and, once secured, its external validity is limited to those other programs whose absence measures and data acquisition methods are similar. Companies' personnel policies and the nature of each work force affect absolute rates of absenteeism and, indeed, how those rates are recorded. Nevertheless, there is a long tradition of using this elusive parameter as an indicator of health promotion program effectiveness.

Absenteeism rates have been measured as an outcome variable in smoking cessation, stress management, hypertension control, health risk assessment, and fitness programs. Program variability and differences in data acquisition notwithstanding, absence rates in general showed a decrease among participants in all of these programs when compared to nonparticipants. Although absolute decreases were modest in many of the work groups, when observed in large employee populations (e.g., school districts) savings reach substantial levels. Certainly the reduction of absenteeism in such employee groups and/or in groups with a high baseline absence rate remains a credible rationale for worksite health promotion programming.

The Impact of Worksite Health Promotion Programs on Productivity

Many companies operating fitness and health promotion programs (or contemplating their introduction) list an increase of productivity as an important anticipated outcome, and some evidence exists to support such an expectation. Use of better productivity measures is needed to quantify such gains.

Improved productivity is an often articulated expectation by managers who are considering implementing worksite health promotion programs. When managers are queried about such expectations after health promotion programs are established, that expectation is diminished. The precise reason for this change has not been determined, but experience suggests that the difficulty in measuring productivity gains may be close to the answer. There is a simple logic to the presumption that employees who are healthier (as a result of their participation in worksite health promotion programs) will be more productive (and utilize fewer health care services and be absent from work less). But just like absenteeism and health care utilization, the factors affecting worker productivity are complex, and health promotion program participation may not result in measurable improvements in productivity, no matter how effective the program.

Shephard, in chapter 9, reviews the rationale for the expectation that health promotion programs will improve productivity and presents the evidence that supports that conclusion. Logic again dictates that workers with improved aerobic fitness, strength, thermoregulatory fitness, and mental health will be more productive workers, and evidence does exist to support those claims. When workers participate in controlled studies in laboratory settings, gains in fitness are generally correlated with improved productivity, with observed increases as high as 20%. However, when such studies are conducted in a worksite setting, in which participation is voluntary and program intensity not as rigorous, gains in workforce productivity drop to much lower levels (1% to 2%).

Productivity-related changes in work satisfaction, feelings of self-efficacy and mood-state, the level of arousal and relaxation of employees, and the adoption of healthy lifestyles have been examined in the context of participation in health promotion programs. Careful examination of these studies suggests that improvements in each of these areas, although present, do not translate into marked changes in productivity. For example, one company's smoking cessation program, while effecting a 12% decrease in smokers, increased company-wide productivity only .22%. Again, we are faced with the dilemma that successful health promotion programs may not produce significant gains in outcome.

Productivity as a measure of health promotion program impact perhaps suffers the most from imprecise measures and poorly designed studies. Lack of control populations, "halo" and Hawthorne effects, and perhaps a bit of "wish bias" all conspire to weaken the credibility of the data (while not diminishing the enthusiasm with which investigators have claimed positive outcomes). Shephard's scholarly and meticulous approach tempers our expectations and provides a sound basis for a conservative evaluation of this topic. Clearly, better indices of productivity must be developed, better research design instituted, and expectations moderated in the study of the relationship of health promotion and productivity.

Health Promotion and Health Benefits Design

Despite questions about its efficacy, changes in benefits design are increasing and will continue to play a greater role in health promotion delivery.

In the early years of the development of worksite health promotion programs, there was a tendency to follow the "field of dreams" philosophy: "Build it and they will come." Many program proponents were subsequently dismayed to find that, unlike legendary baseball players, participants in worksite health promotion did not magically appear upon program implementation. Despite the notion that employer-sponsored health promotion programs and fitness centers were for "their own good," employees did not flock to their doors. Critics continue to claim that these programs attract those already committed to individual responsibility for health, who would otherwise exercise at home or at commercial clubs, and not those less-committed employees who would benefit the most. Are these programs "preaching to the choir," or do health promotion programs truly encourage individuals to change behaviors? While the answer to that question continues to beg improvements in research design and data acquisition, many companies have proactively addressed the issue by providing a variety of incentives and rewards for participation.

Golaszewski, in chapter 10, describes the recent progress made in this area. The logic is simple. Personal employee behavior has a major impact on health care costs. These costs are an ever-increasing burden on company-sponsored health insurance. Punitive "take away" mechanisms only antagonize workers and achieve negligible effects on the rate of health care spending. The use of incentives in general is a well-established business strategy. Thus, incentives to encourage employee participation in health promotion programs could be effective in ultimately lowering costs.

The earliest instance of financial incentives for exercise participation was implemented at Bonnie Bell Cosmetics, and this model continues to be effective. Employees are responsive to financial rewards. Other companies have extended that model to tangible rewards other than money, such as paid travel and additional vacation days. Others still have introduced risk-related insurance plans into their companies as a way of rewarding healthy behaviors. Early evidence suggests that such programs do result in higher participation rates and, presumably, lower health care costs. However, just as with utilization, absenteeism, and productivity, changes in health behaviors in response to incentives is a complex issue and results are mixed thus far.

Since the field is relatively new, few studies are available in the literature. Golaszewski provides the reader with first-hand reports acquired by telephone conversations with managers of incentive-based programs. Five programs are described in detail from around the U.S., ranging in size from large corporations to small, single-campus facilities. In each case, the use of incentives increased program participation, and although the trend was toward a lowering of costs, not all the programs have conducted economic outcome analyses.

Despite a summarizing advocacy for the use of incentives in health promotion programs, caution should be noted in a blanket acceptance of this strategy. Risk-rating insurance plans, for example, ignore the fact that a significant portion of all illness is not risk related. Such plans, therefore, penalize employees for matters over which they have no control and reward others more fortunate, for example, in their selection of parents. Other financial schemes such as returning portions of unused sick days or health care savings accounts to the employees may encourage self-imposed rationing when in fact care should be obtained. Nevertheless, the use of incentives to increase program participation and to improve employee health shows promise, and its evaluation should continue.

Computer Simulation of Health Promotion Program Outcomes

Computer simulations that attempt to model worksite health promotion programs suggest that such programs may be cost beneficial, that economic benefits primarily come from increases in productivity rather than decreases in health care costs, and that program effects vary across different organizations.

Data acquisition remains the most confounding element of program outcome evaluation. To protect employee rights of confidentiality, much of the data are housed in the offices of the human resource director and guarded with the zeal usually ascribed to the trustees of state secrets. In addition, the accuracy of the data is not exacted to that usually obtained for scientific research but is instead subject to the vagaries of self-reporting and incomplete or periodically interrupted collection. Often the need for program outcome evaluation is recognized ex post rather than a priori and the manager is dismayed to find that the required data, potentially available, has simply not been collected. Additionally, there is a cost to data collection and analysis that may not have been budgeted and which often is surprisingly high. For these reasons and perhaps others as well, projections of program outcomes based on normative data and/or logically derived assumptions of program conditions have been generated using computer-driven algorithms, and at relatively low cost.

Terborg, in chapter 11, establishes the rationale for computer-simulated projected outcomes of program evaluation when empirical research is unavailable. The principal utility for corporate decision makers that computer simulations provide is a financial analysis of projected program costs and benefits. Break-even analysis, for example, allows the manager to estimate how many employees must participate in a program for it to recover its costs. Having fewer participants than projected causes the program to lose money; conversely, drawing a greater number than projected brings a positive return on investment. This type of analysis allows the manager to vary the elements of cost to accommodate the actual experience as it unfolds. Indeed, this analysis may be extended to a "base case, best case, and worst case" or "sensitivity" analysis that offers the investigator a range of variable assumptions and resultant outcome projections. Thus, if a program having the minimum number of participants and the lowest benefit derived still offers some positive value to the company, then managers are encouraged to implement the program. If the projection offers little potential for gain, the manager may look to other projects for implementation.

Such analyses were performed on two worksite programs and reported in the literature thus far. One was on a hypothetical organization, and the other projected outcomes for the well-established Coors Brewery health promotion program in Golden, Colorado. In both cases the projections are the inevitable resultant of the assumptions used to drive the program. Sensitivity analyses clearly demonstrated in the first case that some variables had little effect on break-even value (retiree health care costs, medical cost inflation, etc.) while other variables (turnover rates, absenteeism, etc.) had a great impact on the break-even value. Other variables had an intermediate impact (health care costs and participation rates).

In the Coors case, sensitivity analysis suggested that, based on actual participation rates and on partial health care cost and absenteeism data, the program would be a good investment for the company under all but the most marginal conditions. Indeed, this analysis provided the principals at this company with the justification they needed to continue their support.

As the modeling programs become more sophisticated (e.g., "Monte Carlo Analysis"), more and more variable data may be included in the program and a range of outcomes can be projected. With this type of flexibility, program-specific data can be more readily included and the generalizability of the method can be enhanced. Although not a substitute for empirical research, computer modeling offers a real alternative to the health promotion program manager who is unwilling or unable to conduct empirical studies.

Conclusion

The economic impact of worksite health promotion programs is a timely and relevant issue for both health promotion program managers and company managers who must inevitably justify the implementation of these programs to owners. Notwithstanding the difficulty in securing reliable data and in conducting rigorous program evaluations, the industry has generated an impressive number of studies, the collective weight of which supports the conclusion that worksite health promotion programs offer positive financial return on investment. Participants are absent less, spend fewer health care dollars, are more productive, and are injured less frequently than those who do not participate. These programs foster the development of a healthier American work force, with a markedly better quality of life. For the ultimate justification of these programs perhaps that is the real bottom line to which we must point.

References

Fries, J.F., Koop, C.E., Beadle, C.E., Cooper, P.P., England, M.J., Greaves, R.F., Sokolov, J.J., Wright, D., & the Health Project Consortium (1993). Reducing health care costs by reducing the need and demand for medical services. *New England Journal of Medicine*, **329**(5), 321-325.

Pelletier, K.R. (1993). A review and analysis of the health and cost-effectiveness outcome studies of comprehensive health promotion and disease prevention programs at the worksite: 1991-1993 update. *American Journal of Health Promotion*, **8**(1), 50-62.

Warner, K. (1992, September). Keynote address at the Annual Conference of the Association for Worksite Health Promotion (formerly the Association for Fitness in Business), San Diego, CA.

Chapter 3

Noneconomic Benefits of Health Promotion

Steven N. Blair

This chapter will include a relatively brief review of existing studies on health habits and morbidity and mortality from several chronic diseases, as well as studies on changes in health habits and risk of disease. This latter group of studies provides substantial support for the causal role of habits in disease progression and of the importance of health promotion to noneconomic benefits of worksite programs.

Noneconomic Benefits: Providing Definitions and Presenting the Intervention Model

This topic is extremely broad, and a complete coverage of all health promotion activities and all possible outcome measures will not be undertaken. To limit the scope of this report, some definitions are given.

Health Promotion Components

The list of potential health promoting activities or interventions is extensive, perhaps up to two or three dozen topics. The components addressed in this review are those that have been most widely implemented, on which substantial research has been completed, and that have the potential to significantly affect the several outcome measures considered here. The health promotion components included are healthful diet, physical activity, smoking cessation, and weight control.

Outcome Measures

Noneconomic benefits is a vague term and could refer to a very lengthy list of items. The focus taken here is on health and, to a lesser extent, functional capacity. Health status is defined by standard measures of morbidity and mortality as typically used in epidemiological studies. Functional capacity refers to physical fitness parameters such as aerobic power, muscular strength, muscular endurance, and flexibility. The importance of functional capacity relates to the ability to perform routine daily tasks at home, at work, and during leisure time.

Although the focus here is on noneconomic benefits of health promotion, it is obvious that there is an indirect economic benefit to companies if improvements are seen in standard measures of morbidity and mortality and if functional capacity is increased. For example, if a program leads to lower disease rates in a population, health care costs will almost certainly be reduced.

Methods

The approach taken here is not an exhaustive review of all the topics under consideration. Many have a voluminous literature, and there simply is not enough space to offer comprehensive discussions on all topics. Instead, I have summarized the relevant literature to make key points and selected representative studies for inclusion.

The Intervention Model

Chronic diseases are the major cause of death in the industrialized world. At present, more than 1.1 million deaths per year in the United States are due to

- coronary heart disease,
- stroke,
- chronic obstructive pulmonary disease,
- lung cancer,
- cervical cancer,
- breast cancer,
- colorectal cancer,
- cirrhosis, or
- diabetes (Hahn, Teutsch, Rothenberg, & Marks, 1990).

Lifestyle factors are important causes of each of these diseases. The hypothesized link between lifestyle factors and morbidity and mortality from chronic disease is illustrated in Figure 3.1. In order to impact disease endpoints, it is necessary to interrupt the chain of events depicted in the figure. The philosophy of primary prevention focuses attention on the first link, health habits, and this is the orientation taken in this review. It

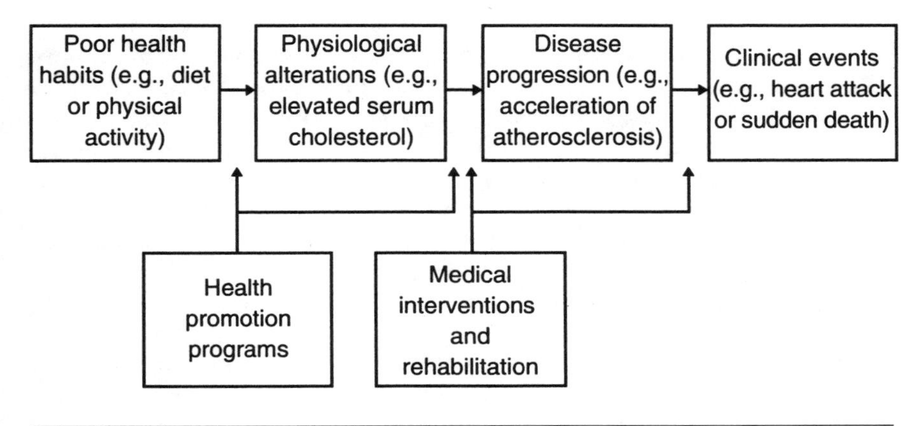

Figure 3.1 Chain of events linking health habits to morbidity and mortality.

is assumed that if health behaviors can be changed, physiological processes will be altered, and disease progression will be delayed or stopped.

It has proven difficult to conduct field studies to show the impact of worksite health promotion on clinical outcomes. Logistical and practical considerations make it unlikely that a health promotion project will be able to trace the effect of health promotion activities through physiological alterations, with tracking of disease processes through their various stages, culminating with lower rates of disease and death. Such studies would require massive sample sizes, a long follow-up interval, and extensive periodic measurements of intermediate variables. A more realistic approach is to focus studies on the various links of the chain. For example, the efficacy of changing health behaviors with worksite health promotion programs can be tested in relatively small studies, and several examples of published reports confirm that well-planned and well-implemented programs result in more employees becoming more physically active or stopping smoking (Blair, Collingwood, et al., 1984; Blair, Piserchia, Wilbur, & Crowder, 1986; Blair, Smith, et al., 1986).

The most firmly established link illustrated in Figure 3.1 is the association between health habits and physiological processes. There is little dispute that cigarette smoking, dietary composition, and physical activity affect such physiological measures as blood lipids, fibrinolytic capacity, insulin sensitivity, and blood pressure.

There has not been much evidence until recently that changes in health habits are associated with disease outcomes. Most of the epidemiological studies on the risk factors for chronic diseases involved a one-time assessment of the behavior with subsequent follow-up for morbidity and mortality. These studies support the hypothesis that poor health habits cause disease, but skeptics can (and do) claim that there may be confounding factors responsible for the observed associations. Tobacco interests have long promulgated the idea that no evidence other than a randomized experiment can provide conclusive support for the hypothesis that smoking causes heart and lung disease. There has never been a single-factor, randomized clinical trial to evaluate the efficacy of changing a health habit to prevent disease. Indeed, such trials are unlikely to be done, due to cost and ethical constraints. For example, would any sensible person recommend that thousands of children be randomized to smoking or nonsmoking groups to determine if coronary heart disease and lung cancer are caused by smoking?

How can we establish the validity of health promotion in disease prevention if randomized trails are not feasible? Fortunately, prospective epidemiological studies in which changes in health behaviors are related to disease outcomes are now becoming available. These studies are a stronger test of the causal hypothesis than the prospective studies with a one-time assessment of health habits. Studies on changes diminish the likelihood that associations between habits and disease are due to genetic factors or confounding variables such as subclinical disease.

Health Habits and Disease

There are numerous health habits linked to disease. In this chapter the presentation is limited to four major modalities: diet, physical activity, weight control, and smoking.

Diet

Few dispute that diet is an important contributor to good health, although some of the issues are complex and consensus has not been attained for several questions. It is generally agreed that a diet high in fruit, vegetables, and whole grains and low in fats (especially saturated fats) and cholesterol is preferred over the typical rich American diet.

Cancer. Several components of the diet have been related to cancer risk. An extensive discussion of all these issues is not possible here, but comprehensive reviews are available (Committee on Diet, Nutrition, and Cancer, 1982). The most important dietary component for cancer risk from a population perspective is fat. High-fat diets have been related to increased risk of the common cancers of the breast, colon, and prostate (Committee on Diet, Nutrition, and Cancer, 1982), although a recent large-scale prospective study weakens the hypothesis that fat increases the risk for breast cancer (Willett et al., 1992).

There is considerable current interest in the role of vitamins A, C, and E in the prevention of certain cancers (Committee on Diet, Nutrition, and

Cancer, 1982; Stampfer et al., 1993; Willett et al., 1990). Higher levels of consumption of these vitamins are associated with reduced cancer risk. Clinical trials are currently under way to further evaluate this hypothesis.

Cardiovascular Diseases. There is little doubt that diet is an important cause of cardiovascular disease. Several decades of research support the theory that diets high in fat (especially saturated fat) and cholesterol cause coronary artery disease (Keys et al., 1986; Neaton et al., 1992), although this relationship may not be as straightforward as previously thought (Jacobs et al., 1992). Cross-cultural studies show a strong direct gradient of coronary heart disease rates across strata of population levels of cholesterol (Keys et al., 1986), and numerous prospective studies demonstrate a high risk of disease for individuals with elevated cholesterol levels (American Heart Association, 1990). There has not been a unifactorial coronary heart disease prevention study with diet, but lowering cholesterol with drugs is associated with lower mortality risk during follow-up (Canner et al., 1986; Frick et al., 1987; Hjermann, Holme, & Leren, 1986; Lipid Research Clinics Program, 1984).

Hypertension is a prevalent disease in the United States and many other industrialized nations. Hypertension has a genetic component but has clear environmental influences as well. Being overweight is strongly related to high blood pressure (National Institutes of Health Consensus Development Panel, 1985; NIH Technology Assessment Panel, 1992; Schotte & Stunkard, 1990; Treatment of Mild Hypertension Research Group, 1991), and some dietary components may also increase risk. There is some evidence that diets high in fat may increase risk of developing hypertension, although consensus on this point has not been reached (National Institutes of Health, 1993). Several cross-cultural studies support the hypothesis that salt intake may increase risk of hypertension. Population blood pressure and salt intake levels are positively related (Intersalt Cooperative Research Group, 1988a), average blood pressures increase in primitive societies as they adopt a Western diet (Intersalt Cooperative Research Group, 1988b), and dietary salt restriction reduces blood pressure (Law, Frost, & Wald, 1991).

Cirrhosis. In the United States, more than 25,000 deaths a year are attributed to cirrhosis, and the rate continues to increase (Hahn et al., 1990). Excessive alcohol intake is the primary dietary component related to cirrhosis. Approximately one third of cirrhosis deaths are estimated to be due to alcohol consumption (Hahn et al., 1990).

Non-Insulin-Dependent Diabetes Mellitus. The major causes of non-insulin-dependent diabetes mellitus are obesity, cigarette smoking, and sedentary lifestyle (Jarrett, 1989). Individual dietary constituents such as sugar and fat intake are probably not associated with increased risk.

Overweight and Obesity. Overweight and obesity are common in the United States (National Institutes of Health, 1985). Although obesity has been considered a major public health problem for decades, we still do not know very much about its causes. Overweight and obesity are clearly caused by positive energy balance over time, and this must be due to too much caloric intake, too little energy expenditure, or both. But these issues remain in question. There is some limited evidence that diets high in fat may predispose an individual to obesity, but the data are not conclusive. No other dietary components, other than total caloric intake, are likely to be causes of overweight and obesity.

Physical Activity

Physical inactivity increases the risk for several health problems and diseases (Blair, Kohl, Gordon, & Paffenbarger, 1992). Humans evolved as active animals—first as scavengers and later as hunter-gatherers. It is only within the past 200 years with the advent of the Industrial Revolution and increased reliance on fossil fuels for energy that large numbers of individuals have come to lead a sedentary existence. This unnatural way of life almost certainly plays a role in the development of modern health problems.

Cardiovascular Diseases. There have been numerous reviews of the relation of physical inactivity to cardiovascular diseases; an excellent example is the one by Powell, Thompson, Caspersen, and Kendrick (1987). Further, I recently summarized the epidemiological studies on physical activity or physical fitness and cardiovascular disease published since 1987 for the International Conference on Physical Activity, Fitness, and Health held in Toronto in 1992 (Blair, 1994). Table 3.1 presents a summary of these recent studies. These studies had excellent epidemiological methods, good assessment of activity or fitness, and a relatively complete surveillance system for detecting mortality, and they provide strong support for the causal role of physical inactivity in the development of cardiovascular disease. The American Heart Association published a revised Statement on Exercise in July 1992 in which inactivity is designated as the fourth major risk factor for coronary artery disease, joining hypertension, high blood cholesterol, and cigarette smoking (Fletcher et al., 1992).

Although the evidence is not as extensive or persuasive as it is for coronary heart disease, other cardiovascular diseases are also associated with sedentary living habits. Risk of stroke appears to be higher in inactive individuals (Kohl & McKenzie, 1994), and low levels of activity and fitness are associated with approximately a 30% increased risk of developing physician-diagnosed hypertension (Blair, Goodyear, Gibbons, & Cooper, 1984; Paffenbarger, Wing, Hyde, & Jung, 1983).

Table 3.1
Summary of Studies on Physical Activity or Physical Fitness and Health, 1987 to 1992

Study group	Design	Physical activity/ fitness measure	Endpoint	Main results
Study group				
1. Representative sample, Alameda County, CA; 4,174 men and women aged 38 years or more at baseline (Kaplan, Seeman, Cohen, Knudsen, & Guralik, 1987)	Prospective, 17-year follow-up	Questionnaire, leisure-time activity	All-cause mortality, 1,219 deaths	RR = 1.38[a] (1.17, 1.62)[b], low/high activity
2. High-risk men in the Multiple Risk Factor Intervention Trial; 12,138 men aged 35–57 years at baseline (Leon, Connett, Jacobs, & Rauramaa, 1987)	Prospective, 7-year follow-up	Questionnaire, leisure-time activity	CHD death, 225 deaths	RR = 0.63[a] (0.43, 0.86) 2nd/1st tertile; RR = 0.64[a] (0.47, 0.88) 3rd/1st tertile
3. Seventh-Day Adventist men; 9,484 men aged 30 years and older at baseline (Lindsted, Tonstad, & Kuzma, 1991)	Prospective, 26-year follow-up	Questionnaire, leisure and work activity	CHD death, 1,351 deaths	Age at crossover (RR = 1.0), 83.9[a] (74.9, 92.8) for moderate/low; 76.5[a] (66.0, 86.9) for high/low
4. British Civil Servants; 9,376 men aged 45 to 64 years at baseline (Morris, Clayton, Everitt, Semmence, & Burgess, 1990)	Prospective, 9-year follow-up	Questionnaire, leisure-time activity	CHD death, 289 deaths	RR = 0.34[c] (0.18, 0.66) for frequent vigorous aerobic exercise (VE)/no VE

5. Population-based study in southwestern Finland; 636 men aged 40 to 59 years at baseline (Pekkanen et al., 1987)	Prospective, 20-year follow-up	Questionnaire, leisure and work activity	CHD death, 106 deaths	RR = 1.3[a], low/high activity
6. Random population samples from eastern Finland; 15,088 men and women aged 30 to 59 years at baseline (Salonen, Slater, Tuomilehto, & Rauramaa, 1988)	Prospective, 6-year follow-up	Questionnaire, leisure and work activity	CHD death, 102 deaths (90 men, 12 women)	RR = 1.4[a] (1.1, 1.7), low (either leisure or work)/high activity
7. U.S. railroad workers; 2,548 white men (Slattery, Jacobs, & Nichaman, 1989)	Prospective, 17- to 20-year follow-up	Questionnaire, leisure-time activity	CHD death	RR = 1.28[a] (0.99, 1.63), sedentary/active
8. Japanese-American men in Honolulu; 7,644 men aged 45 to 69 years at baseline (Donahue, Abbott, Reed, & Yano, 1988)	Prospective, 12-year follow-up	Questionnaire, estimate of total energy expenditure	Fatal and nonfatal CHD, 444 events	RR = 0.69[c] (0.54, 0.88), in active vs. inactive men 45–64 years; RR = 0.42[c] (0.18, 0.96), in active vs. inactive men 65 years and older
9. British Regional Heart Study; 5,714 men aged 40 to 59 years, free of CHD at baseline (Shaper & Wannamethee, 1991)	Prospective, 8-year follow-up	Questionnaire, leisure-time activity	Fatal (N = 217) and nonfatal (N = 271) CHD	Men who reported moderate or moderately vigorous activity "experienced less than half the rate seen in inactive men."
Physical fitness				
1. Cooper Clinic patients; 3,120 women, 10,224 men (Blair et al., 1989)	Prospective, $\bar{x}$ follow-up of slightly more than 8 years	Maximal exercise tolerance, treadmill test	Cardiovascular disease (CVD) death, 73 deaths (7 women, 66 men)	RR = 9.25[c] (–5.1, 0.5)[d] for women; RR = 7.93[c] (–8.8, –3.3)[d] for men, low/high fitness

(continued)

Table 3.1 *(continued)*

Study group	Design	Physical activity/ fitness measure	Endpoint	Main results
2. Lipid Research Clinics study; 3,106 men, 30 to 69 years at baseline (Ekelund et al., 1988)	Prospective, 8.5-year follow-up	Submaximal exercise test, treadmill	CHD death, CVD death, 45 deaths	RR = 6.5 (1.5, 28.7) for CHD death; RR = 8.5 (2.0, 36.7) for CVD death, least fit quartile/most fit quartile; adjusted RR = 2.8[a] (1.3, 6.1) for CHD; adjusted RR = 3.6[a] (1.6, 5.6) for CVD
3. Cooper Clinic patients; 2,926 men who were hypertensive at baseline (Blair, Kohl, Barlow, & Gibbons, 1991)	Prospective, $\bar{x}$ follow-up of slightly more than 8 years	Maximal exercise tolerance, treadmill test	CVD death, 63 deaths	RR = 2.3[c] (1.3, 3.8), low/high fitness
4. Company or government employees in Oslo; 2,014 men, 40 to 59 years at baseline (Lie, Mundal, & Erikssen, 1985)	Prospective, 7-year follow-up	Submaximal exercise test, cycle ergometer	CHD death, 58 deaths	RR ≈ 4.8, least fit quartile/most fit quartile
5. U.S. railroad workers; 2,431 white men, 22 to 79 years at baseline (Slattery & Jacobs, 1988)	Prospective, $\bar{x}$ follow-up of approximately 20 years	Submaximal exercise test, treadmill	CHD and CVD death, 258 CHD deaths	RR = 1.45[c] for CHD and 1.51 for CVD; adjusted RR = 1.20[a] (1.10, 1.26) for CHD; low/high fitness

[a]Relative risk (RR) adjusted for age and other major CHD risk factors. [b]Numbers in parentheses are 95% confidence intervals. [c]Relative risk adjusted for age only, although additional adjustments for other factors had a negligible effect. [d]95% confidence interval for linear trend slope.
Note. From Blair, 1994. Adapted by permission.

Cancer. Physical inactivity is associated with an increased risk for certain cancers. The relation of sedentary habits to cancer risk has been reviewed (Kohl, LaPorte, & Blair, 1988; Lee, 1994). The greatest amount of data on physical activity and cancer risk is for colon cancer, where an approximate doubling in risk is noted for the most sedentary, when compared with the most active individuals. There also is evidence supporting increased risk for lung, breast, and prostate cancers in inactive persons (Lee, 1994). The recent review of Lee for the International Conference on Physical Activity, Fitness, and Health is up-to-date and thorough and is highly recommended for those who want further detail on this topic.

Non-Insulin-Dependent Diabetes Mellitus. Only a few years ago, most experts would say that inactivity may increase the risk of developing non-insulin-dependent diabetes mellitus but that hard evidence was lacking. This situation has recently changed. Several large prospective studies with excellent methods demonstrate up to a 35% reduced risk of non-insulin-dependent diabetes mellitus in active men and women when compared with their sedentary peers (Helmrich, Ragland, Leung, & Paffenbarger, 1991; Manson et al., 1991, 1992). When the epidemiological evidence is considered along with the physiological data from controlled exercise training studies, it now seems reasonable to conclude that activity may help prevent the development of non-insulin-dependent diabetes mellitus.

Overweight. The causes of overweight remain obscure, as previously mentioned, although sedentary habits logically appear to play a role. Cross-sectional studies are nearly unanimous in showing that active and fit men and women weigh less than sedentary and unfit individuals (Cooper et al., 1976; Gibbons, Blair, Cooper, & Smith, 1983). Exercise intervention contributes to weight loss, but the additional amount of weight lost when exercise is added to dieting is not large (on the order of two additional kilograms) (King, Frey-Hewitt, Dreon, & Wood, 1989; Wood, Stefanick, Williams, & Haskell, 1991; Wood et al., 1988). It appears that the protein sparing modified fasting approach with liquid formula diets will cause greater and more rapid weight loss than can be achieved by exercise (Kohl et al., in review). Physical activity appears to play an important role in maintaining weight lost in an intervention program (Pavlou, Steffee, Lerman, & Burrows, 1985), and numerous other benefits accrue to the active overweight—or formerly overweight—individual (Blair, 1993).

The role of a physically active lifestyle in the prevention of obesity has not been thoroughly investigated. Some studies show that inactive individuals are no more likely to gain weight during follow-up than active individuals (Klesges, Klesges, Haddock, & Eck, 1992), and some studies suggest a benefit for activity (Williamson et al., 1993). This issue needs much additional research.

Functional Capacity. Physical activity increases physical fitness and various aspects of physical function. Many well-controlled exercise training studies over the past several decades have elucidated the nature of the dose–response relation between activity and fitness (American College of Sports Medicine, 1991; American College of Sports Medicine Position Stand, 1990). Earlier studies tended to focus on improving maximal aerobic power in younger and middle-aged individuals. Recent work has expanded this concept to other types of fitness and has included older individuals. Life in modern society does not demand a high level of physical fitness to meet the needs of most daily occupational, leisure, and household tasks. The majority of younger and middle-aged men and women have sufficient reserves of function to meet these demands, even if they actually are rather sedentary in their habits. After years and decades of progressive decline, however, lack of fitness may begin to impair performance so that some individuals cannot meet even the minimal fitness requirements of daily living. The prevalence of this relative disability increases to substantial levels in older men and women (Cornoni-Huntley, Brock, Ostfeld, Taylor, & Wallace, 1986). Recent follow-up studies suggest that inactive individuals are more likely to become impaired as they age than are active persons (Mor et al., 1989). Furthermore, strength training studies in men and women 90 years of age and older show that impressive gains in strength are possible in this group and that higher levels of strength reduce disability (Fiatarone et al., 1990). One of the major benefits of an active lifestyle is the concomitant enhanced function that results.

Weight Control

Overweight is considered a major public health problem in the United States (U.S. Department of Health and Human Services, 1991). Much attention is paid to this issue by health professionals, but the public is probably even more concerned. Nearly one half of adult American women and one quarter of American men claim to be dieting. The weight loss industry is large, with billions of dollars spent annually in the pursuit of thinness (NIH Technology Assessment Conference Panel, 1992).

A body mass index (BMI) of 27 kg/m^2 is considered the upper limit of a healthful body composition (National Institutes of Health, 1985), and nearly one quarter of adults in the United States exceed that value. Overweight is associated with several health problems including coronary heart disease, stroke, non-insulin-dependent diabetes mellitus, hypertension, and arthritis (National Institutes of Health, 1985; Després et al., 1990); but weight loss in previously overweight individuals has not been conclusively shown to reduce risk. In fact, some recent reports suggest that weight loss may actually be harmful. We evaluated the association of weight change and mortality in 10,529 high-risk men who participated in the Multiple Risk Factor Intervention Trial (Blair, Shaten, Brownell,

Collins, & Lissner, 1993). Weight change was determined over the 6 to 7 years of the intervention phase of the study, with subsequent mortality surveillance for an additional 3.8 years. Men in the middle tertile of baseline BMI (26.09 to 28.82 kg/m^2) who lost 5% or more of their baseline weight (average 5% loss = approximately 4 kg) had a nearly three-fold increased risk for cardiovascular disease mortality and a two-fold increased risk for all-cause mortality when compared with the men whose weights never varied by as much as 5% of their baseline weights. Thus, although we know that weight loss causes an improvement in clinical status (lower blood pressure and cholesterol), we do not have good evidence that morbidity and mortality risk is reduced by weight loss. Additional studies must be conducted to resolve this paradox.

Cigarette Smoking

Cigarette smoking causes nearly 1,000 deaths a day in the United States due to its major contribution to the nine chronic diseases listed earlier in this report (Hahn et al., 1990). The total (361,911 chronic disease deaths per year) exceeds those for the other major lifestyle risk factors reviewed here, each of which cause about 250,000 chronic disease deaths per year in the United States. Cigarette smoking increases the risk for most of the diseases mentioned in this chapter, although its major impact on mortality statistics is via cardiovascular disease and lung cancer (U.S. Department of Health and Human Services, 1989). Virtually all public health authorities, clinicians, and other health scientists accept the hypothesis that cigarette smoking causes higher rates of morbidity and mortality. There have been literally thousands of studies from all three main branches of medical science—epidemiology, pathology, and experiments—that confirm the causal hypothesis (U.S. Department of Health and Human Services, 1989). In addition, we have known for many years that stopping smoking is associated with a return to lower risk status (Paffenbarger et al., 1992), almost immediately for coronary heart disease and after several years for lung cancer.

I believe that the harmful effects of smoking and the beneficial effects of quitting have been proven and will not review these issues further here. It is sufficient to state that cigarette smoking is hazardous and that any health promotion program must have smoking control as one of the objectives.

Studies on Changes in Health Habits

The reviewed evidence clearly indicates strong associations between lifestyle characteristics and risk of chronic disease. These data provide a substantial rationale for implementing broadly based health promotion programs. Of course associations between risk factors and disease do not prove causality, and additional information is needed. Because large-scale, randomized experiments to evaluate the impact of changing health habits on morbidity and mortality are not likely to be done, well-designed prospective observational studies of health behavior change are needed to evaluate the hypothesis that these changes are beneficial to health. Such studies are now available.

The Alameda County Study

Investigators at the Human Population Laboratory, a part of the California Department of Health Services, began a long-term observational study in 1965 in a representative population sample from Alameda County, California (Kaplan & Haan, 1989). In 1974, a second survey was completed in 863 men and women who were 50 years of age or older at baseline. Mortality follow-up was maintained in the cohort from 1974 to 1983. Individuals who quit smoking in the interval 1965 to 1974 had a reduced risk of all-cause mortality compared with individuals who continued to smoke, and those who increased their participation in physical activity were at lower risk than those who did not. These analyses support the hypothesis that health behavior changes in men and women 50 years of age and older provide benefits.

College Alumni Studies

Paffenbarger and colleagues have followed large populations of college alumni from the University of Pennsylvania and Harvard College for several years. Social characteristics, health and medical histories, and health habits have been assessed on several occasions by mail questionnaire surveys in both populations.

University of Pennsylvania. In 1962 and again in 1976, Paffenbarger surveyed 6,242 alumni for smoking status and physical activity habits (Paffenbarger et al., 1992). About 61% of the men were nonsmokers in 1962 and 39% were smokers. By 1976, only 18% of the smokers were still smoking and 21% had quit since the 1962 survey. Approximately 64% of the men reported no vigorous physical activity in 1962, but 14% of them had taken up vigorous activity by 1976. The effect of stopping smoking and starting a vigorous exercise program on cardiovascular disease mortality in these men is shown in Figure 3.2. Substantial reductions of approximately 50% in the age-adjusted death rate support the desirability of making changes in smoking and exercise behavior. A striking feature of

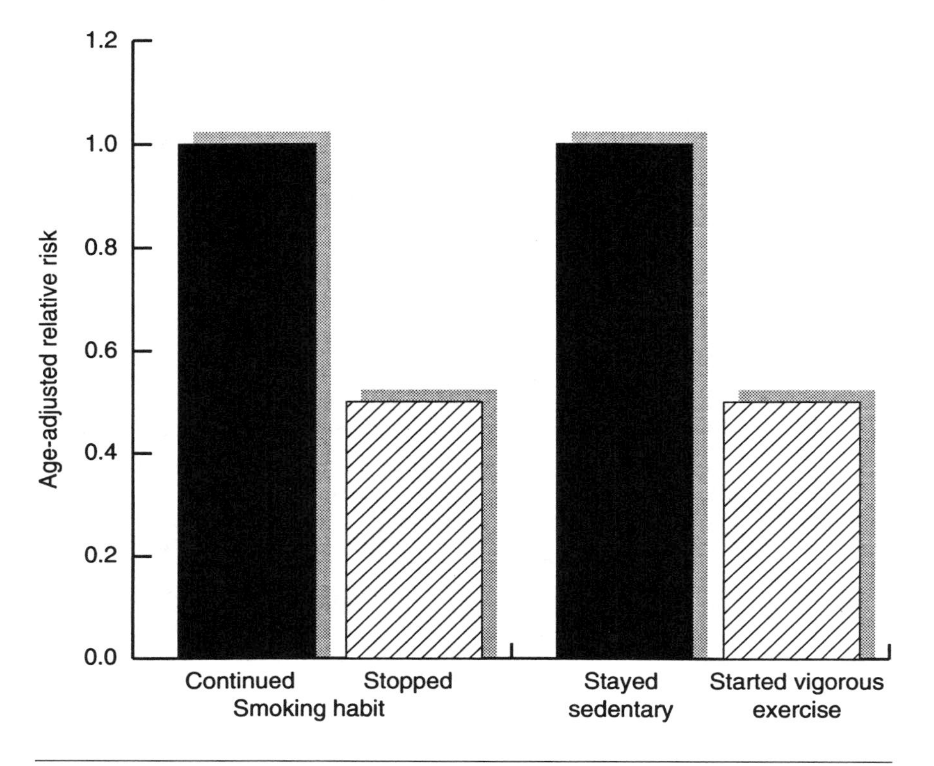

Figure 3.2 Risk of cardiovascular disease death by changes in smoking and exercise habits for 6,242 University of Pennsylvania alumni. The vertical axis shows the reduction in risk associated with stopping smoking and starting vigorous exercise.
Note. Data from "Influence of Cigarette Smoking and Its Cessation on the Risk of Cardiovascular Disease" by R.S. Paffenbarger, Jr. et al. In *Control of Tobacco-related Cancers and Other Diseases. International Symposium, 1990* by P.C. Gupta, J.E. Hamner, III, and P.R. Murti (Eds.), 1992, Bombay: Oxford University Press.

these data is that starting to exercise appears to be as important as stopping smoking in terms of reducing cardiovascular disease risk.

Harvard College. Paffenbarger et al. recently reported the impact of changes in lifestyle characteristics on mortality in Harvard alumni (Paffenbarger et al., 1993). There were 10,269 alumni who completed a baseline questionnaire in either 1962 or 1966 and who subsequently responded to a second questionnaire in 1977. The cohort was followed for mortality from 1977 to 1985, with 476 deaths occurring in that interval. Changes in participation in moderately vigorous sports (at least 4.5 times resting energy expenditure), cigarette smoking, hypertension, and overweight are presented in Figure 3.3 relative to the risk of coronary heart disease

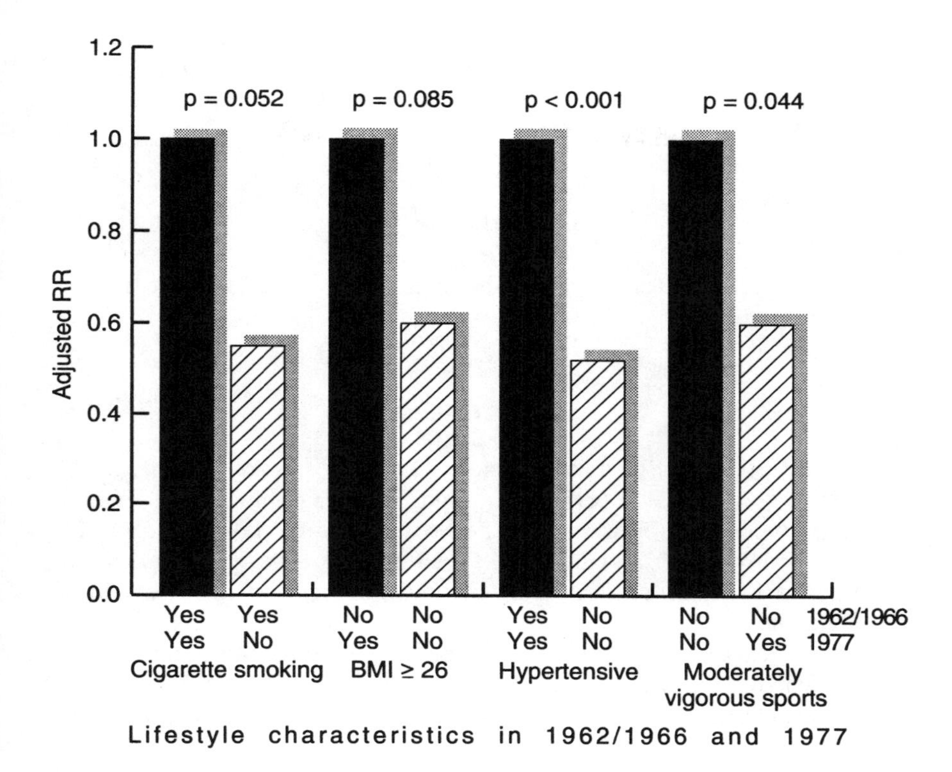

Figure 3.3 Adjusted relative risk (RR) for coronary heart disease by lifestyle changes for 10,269 Harvard alumni. The vertical axis shows the reduction in risk associated with favorable changes in cigarette smoking, body mass index, hypertension, and moderately vigorous sports play from 1962 or 1966 to 1977. *Note.* Data from "The Association of Changes in Physical-Activity Level and Other Lifestyle Characteristics With Mortality Among Men" by R.S. Paffenbarger, Jr. et al., 1993, *New England Journal of Medicine*, **328**, pp. 538-545.

mortality. These relative risks (RR) are adjusted for age and each of the other variables in the figure. The reductions in risk are comparable (approximately 45%) for each of the characteristics.

Summary

Several key health promotion modalities and their association to noneconomic health benefits have been reviewed in this chapter. Health benefits are defined here as delay or avoidance of morbidity or mortality from chronic diseases. There are strong scientific foundations for the hypothesis that these health promotion modalities are causally related to the incidence of chronic disease and mortality from these problems. Recent studies showing important reductions in mortality risk in middle-aged and older individuals who make positive health behavior changes provide significant additional information. I conclude that health promotion leads to substantial noneconomic benefits.

Acknowledgment

I thank Laura Becker for preparing the manuscript, producing the figures, and proofreading.

References

American College of Sports Medicine (1991). *Guidelines for graded exercise testing and exercise prescription* (4th ed.). Philadelphia: Lea & Febiger.

American College of Sports Medicine Position Stand (1990). The recommended quantity and quality of exercise for developing and maintaining cardiorespiratory and muscular fitness in healthy adults. *Medicine and Science in Sports and Exercise*, **22**, 265-274.

American Heart Association, Task Force on Cholesterol Issues (1990). The cholesterol facts: A summary of the evidence relating dietary fats, serum cholesterol, and coronary heart disease. *Circulation*, **81**, 1721-1733.

Blair, S.N. (1993). Evidence for success of exercise in weight loss and control. *Annals of Internal Medicine*, **119**, 702-706.

Blair, S.N. (1994). Physical activity, fitness, and coronary heart disease. In C. Bouchard, R.J. Shephard, & T. Stephens (Eds.), *Physical activity, fitness, and health* (pp. 579-589). Champaign, IL: Human Kinetics.

Blair, S.N., Collingwood, T.R., Reynolds, R., Smith, M., Hagan, D., & Sterling, C.L. (1984). Health promotion for educators: Impact on health behaviors, satisfaction, and general well-being. *American Journal of Public Health*, **74**, 147-149.

Blair, S.N., Goodyear, N.N., Gibbons, L.W., & Cooper, K.H. (1984). Physical fitness and incidence of hypertension in healthy normotensive men and women. *Journal of the American Medical Association*, **252**, 487-490.

Blair, S.N., Kohl, III, H.W., Barlow, C.E., & Gibbons, L.W. (1991). Physical fitness and all-cause mortality in hypertensive men. *Annals of Medicine*, **23**, 307-312.

Blair, S.N., Kohl, H.W., III, Gordon, N.F., & Paffenbarger, Jr., R.S. (1992). How much physical activity is good for health? *Annual Review of Public Health*, **13**, 99-126.

Blair, S.N., Kohl, H.W., III, Paffenbarger, Jr., R.S., Clark, D.G., Cooper, K.H., & Gibbons, L.W. (1989). Physical fitness and all-cause mortality:

A prospective study of healthy men and women. *Journal of the American Medical Association*, **262**, 2395-2401.

Blair, S.N., Piserchia, P.V., Wilbur, C.S., & Crowder, J.H. (1986). A public health intervention model for work-site health promotion: Impact on exercise and physical fitness in a health promotion plan after 24 months. *Journal of the American Medical Association*, **255**, 921-926.

Blair, S.N., Shaten, J., Brownell, K., Collins, G., & Lissner, L. (1993). Body weight change, all-cause mortality, and cause-specific mortality in the Multiple Risk Factor Intervention Trial. *Annals of Internal Medicine*, **119**, 749-757.

Blair, S.N., Smith, M., Collingwood, T.R., Reynolds, R., Prentice, M.C., & Sterling, C.L. (1986). *Preventive Medicine*, **15**, 166-175.

Canner, P.L., Berge, K.G., Wenger, N.K., Stamler, J., Friedman, L., Prineas, R.J., & Friedwald, W. (1986). Fifteen year mortality in Coronary Drug Project patients: Long-term benefit with niacin. *Journal of the American College of Cardiology*, **8**, 1245-1255.

Committee on Diet, Nutrition, and Cancer (1982). *Diet, nutrition, and cancer*. Washington, DC: National Academy Press.

Cooper, K.H., Pollock, M.L., Martin, R.P., White, S.R., Linnerud, A.C., & Jackson, A. (1976). Physical fitness levels vs selected coronary risk factors. *Journal of the American Medical Association*, **236**, 166-169.

Cornoni-Huntley, J., Brock, D.B., Ostfeld, A.M., Taylor, J.O., & Wallace, R.B. (Eds.) (1986). *Established populations for epidemiologic studies of the elderly*. Bethesda, MD: National Institutes of Health.

Després, J-P., Moorjani, S., Lupien, P.J., Tremblay, A., Nadeau, A., & Bouchard, C. (1990). Regional distribution of body fat, plasma lipoproteins, and cardiovascular disease. *Arteriosclerosis*, **10**, 497-511.

Donahue, R.P., Abbott, R.D., Reed, D.M., & Yano, K. (1988). Physical activity and coronary heart disease in middle-aged and elderly men: The Honolulu Heart Program. *American Journal of Public Health*, **78**, 683-685.

Ekelund, L-G., Haskell, W.L., Johnson, J.L., Whaley, F.S., Criqui, M.H., & Sheps, D.S. (1988). Physical fitness as a predictor of cardiovascular mortality in asymptomatic North American men: The Lipid Research Clinics Mortality Follow-up Study. *New England Journal of Medicine*, **319**, 1379-1384.

Fiatarone, M.A., Marks, E.C., Ryan, N.D., Meredith, C.N., Lipsitz, L.A., & Evans, W.J. (1990). High-intensity strength training in nonagenarians: Effects on skeletal muscle. *Journal of the American Medical Association*, **263**, 3029-3034.

Fletcher, G.F., Blair, S.N., Blumenthal, J., Caspersen, C., Chaitman, B., Epstein, S., Falls, H., Sivarajan-Froelicher, E.S., Froelicher, V.F., & Pina, I.L. (1992). Benefits and recommendations for physical activity programs for all Americans: A statement for health professionals by the Committee on Exercise and Cardiac Rehabilitation of the Council on Clinical Cardiology, American Heart Association. *Circulation*, **86**, 340-344.

Frick, M.H., Elo, O., Haapa, K., Heinonen, O.P., Heinsalmi, P., Helo, P., Huttunen, J.K., Kaitaniemi, P., Koskinen, P., & Manninen, V. (1987). Helsinki Heart Study: Primary-prevention trial with gemfibrozil in middle-aged men with dyslipidemia: Safety in treatment, changes in risk factors, and incidence of coronary heart disease. *New England Journal of Medicine*, **317**, 1237-1245.

Gibbons, L.W., Blair, S.N., Cooper, K.H., & Smith, M. (1983). Association between coronary heart disease risk factors and physical fitness in healthy adult women. *Circulation*, **67**, 977-983.

Hahn, R.A., Teutsch, S.M., Rothenberg, R.B., & Marks, J.S. (1990). Excess deaths from nine chronic diseases in the United States, 1986. *Journal of the American Medical Association*, **264**, 2654-2659.

Helmrich, S.P., Ragland, D.R., Leung, R.W., & Paffenbarger, R.S., Jr. (1991). Physical activity and reduced occurrence of non-insulin-dependent diabetes mellitus. *New England Journal of Medicine*, **325**, 147-152.

Hjermann, I., Holme, I., & Leren, P. (1986). Oslo Study Diet and Antismoking Trial. Results after 102 months. *American Journal of Medicine*, **80**, 7-11.

Intersalt Cooperative Research Group (1988a). Sodium, potassium, body mass, alcohol and blood pressure: The INTERSALT Study. *Journal of Hypertension*, **6**(Suppl. 4), S584-586.

Intersalt Cooperative Research Group (1988b). Intersalt: An international study of electrolyte excretion and blood pressure. Results for 24 hour urinary sodium and potassium excretion. *British Medical Journal*, **297**, 319-328.

Jacobs, D., Blackburn, H., Higgins, M., Reed, D., Iso, H., McMillan, G., Neaton, J., Nelson, J., Potter, J., Rifkind, B., Rossouw, J., Shekelle, R., & Yusuf, S. (1992). Report of the Conference on Low Blood Cholesterol: Mortality associations. *Circulation*, **86**, 1046-1060.

Jarrett, R.J. (1989). Epidemiology and public health aspects of non-insulin-dependent diabetes mellitus. *Epidemiologic Reviews*, **11**, 151-171.

Kaplan, G.A., & Haan, M.N. (1989). Is there a role for prevention among the elderly? Epidemiological evidence from the Alameda County Study. In M.G. Ory & K. Bond (Eds.), *Aging and health care: Social science and policy perspectives*. London: Tavistock Publications.

Kaplan, G.A., Seeman, T.E., Cohen, R.D., Knudsen, L.P., & Guralnik, J. (1987). Mortality among the elderly in the Alameda County Study: Behavioral and demographic risk factors. *American Journal of Public Health*, **77**, 307-312.

Keys, A., Menotti, A., Karvonen, M.J., Arvanis, C., Blackburn, H., Buzina, R., Djordjevic, B.S., Dontas, A.S., Fidanza, F., & Keys, M.H. (1986). The diet and 15-year death rate in the seven countries study. *American Journal of Epidemiology*, **124**, 903-915.

King, A.C., Frey-Hewitt, B., Dreon, D.M., & Wood, P.D. (1989). Diet vs. exercise in weight maintenance: The effects of minimal intervention strategies on long-term outcomes in men. *Archives of Internal Medicine*, **149**, 2741-2746.

Klesges, R.C., Klesges, L.M., Haddock, C.K., & Eck, L.H. (1992). A longitudinal analysis of the impact of dietary intake and physical activity on weight change in adults. *American Journal of Clinical Nutrition*, **55**, 818-822.

Kohl, H.W., Hoerr, R.A., Nichaman, M.Z., Bunker, A., McPherson, R.S., & Blair, S.N. (in review). Short-term weight loss efficacy of a very low calorie diet-based weight loss program. A study of 3,352 women and men.

Kohl, H.W., LaPorte, R.E., & Blair, S.N. (1988). Physical activity and cancer: An epidemiological perspective. *Sports Medicine*, **6**, 222-237.

Kohl, H.W., III, & McKenzie, J.D. (1994). Physical activity, physical fitness, and stroke. In C. Bouchard, R.J. Shephard, & T. Stephens (Eds.), *Physical activity, fitness, and health* (pp. 609-620). Champaign, IL: Human Kinetics.

Law, M.R., Frost, C.D., & Wald, N.J. (1991). By how much does dietary salt reduction lower blood pressure? I—Analysis of observational data among populations. *British Medical Journal*, **302**, 811-815.

Lee, I-M. (1994). Physical activity, fitness, and cancer. In C. Bouchard, R.J. Shephard, & T. Stephens (Eds.), *Physical activity, fitness, and health* (pp. 814-826). Champaign, IL: Human Kinetics.

Leon, A.S., Connett, J., Jacobs, Jr., D.R., & Rauramaa, R. (1987). Leisure-time physical activity levels and risk of coronary heart disease and death: The Multiple Risk Factor Intervention Trial. *Journal of the American Medical Association*, **258**, 2388-2395.

Lie, H., Mundal, R., & Erikssen, J. (1985). Coronary risk factors and incidence of coronary death in relation to physical fitness. Seven-year follow-up study of middle-aged and elderly men. *European Heart Journal*, **6**, 147-157.

Lindsted, K.D., Tonstad, S., & Kuzma, J. (1991). Self-report of physical activity and patterns of mortality in Seventh-day Adventist men. *Journal of Clinical Epidemiology*, **44**, 355-364.

Lipid Research Clinics Program (1984). The Lipid Research Clinics Coronary Primary Prevention Trial results. I. Reduction in incidence of coronary heart disease. *Journal of the American Medical Association*, **251**, 351-364.

Manson, J.E., Nathan, D.M., Krolewski, A.S., Stampfer, M.J., Willett, W.C., & Hennekens, C.H. (1992). A prospective study of exercise and incidence of diabetes among U.S. male physicians. *Journal of the American Medical Association*, **268**, 63-67.

Manson, J.E., Rimm, E.B., Stampfer, M.J., Colditz, G.A., Willett, W.C., Krolewski, A.S., Rosner, B., Hennekens, C.H., & Speizer, F.E. (1991).

Physical activity and incidence of non-insulin-dependent diabetes mellitus in women. *Lancet*, **338**, 774-778.
Mor, V., Murphy, J., Masterson-Allen, S., Willey, C., Razmpour, A., Jackson, M.E., Greer, D., & Katz, S. (1989). Risk of functional decline among well elders. *Journal of Clinical Epidemiology*, **42**, 895-904.
Morris, J.N., Clayton, D.G., Everitt, M.G., Semmence, A.M., & Burgess, E.H. (1990). Exercise in leisure time: Coronary attack and death rates. *British Heart Journal*, **63**, 325-334.
National Institutes of Health (1993). *The fifth report of the Joint National Committee on Detection, Evaluation, and Treatment of High Blood Pressure* (NIH Publication No. 93-1088). Washington, DC: U.S. Government Printing Office.
National Institutes of Health Consensus Development Panel on the Health Implications of Obesity (1985). Health implications of obesity: National Institutes of Health Consensus Development Conference statement. *Annals of Internal Medicine*, **103**, 1073-1077.
National Institutes of Health Technology Assessment Conference Panel (1992). Methods for voluntary weight loss and control. *Annals of Internal Medicine*, **116**, 942-949.
Neaton, J.D., Blackburn, H., Jacobs, D., Kuller, L., Lee, D-J., Sherwin, R., Shih, J., Stamler, J., & Wentworth, D. (1992). Serum cholesterol level and mortality findings for men screened in the Multiple Risk Factor Intervention Trial. *Archives of Internal Medicine*, **152**, 1490-1500.
Paffenbarger, Jr., R.S., Hyde, R.T., Leung, R.W., Jung, D.L., Wing, A.L., & Hsieh, C. (1992). Influence of cigarette smoking and its cessation on the risk of cardiovascular disease. In P.C. Gupta, J.E. Hamner, III, & P.R. Murti (Eds.), *Control of Tobacco-related Cancers and Other Diseases. International Symposium, 1990*. Bombay: Oxford University Press.
Paffenbarger, Jr., R.S., Hyde, R.T., Wing, A.L., Lee, I-M., Jung, D.L., & Kampert, J.B. (1993). The association of changes in physical-activity level and other lifestyle characteristics with mortality among men. *New England Journal of Medicine*, **328**, 538-545.
Paffenbarger, Jr., R.S., Wing, A.L., Hyde, R.T., & Jung, D.L. (1983). Physical activity and incidence of hypertension in college alumni. *American Journal of Epidemiology*, **117**, 245-257.
Pavlou, K., Steffee, W., Lerman, R., & Burrows, B. (1985). Effects of dieting and exercise on lean body mass, oxygen uptake, and strength. *Medicine and Science in Sports and Exercise*, **17**, 466-471.
Pekkanen, J., Marti, B., Nissinen, A., Tuomilehto, J., Punsar, S., & Karvonen, M.J. (1987). Reduction of premature mortality by high physical activity: A 20-year follow-up of middle-aged Finnish men. *Lancet*, **i**, 1473-1477.
Powell, K.E., Thompson, P.D., Caspersen, C.J., & Kendrick, J.S. (1987). Physical activity and the incidence of coronary heart disease. *Annual Review of Public Health*, **8**, 253-287.

Salonen, J.T., Slater, J.S., Tuomilehto, J., & Rauramaa, R. (1988). Leisure time and occupational physical activity: Risk of death from ischemic heart disease. *American Journal of Epidemiology*, **127**, 87-94.

Schotte, D.E., & Stunkard, A.J. (1990). The effects of weight reduction on blood pressure in 301 obese patients. *Archives of Internal Medicine*, **150**, 1701-1704.

Shaper, A.G., & Wannamethee, G. (1991). Physical activity and ischaemic heart disease in middle-aged British men. *British Medical Journal*, **66**, 384-394.

Slattery, M.L., & Jacobs, Jr., D.R. (1988). Physical fitness and cardiovascular disease mortality: The U.S. Railroad Study. *American Journal of Epidemiology*, **127**, 571-580.

Slattery, M.L., Jacobs, Jr., D.R., & Nichaman, M.Z. (1989). Leisure time physical activity and coronary heart disease death: The U.S. Railroad Study. *Circulation*, **79**, 304-311.

Stampfer, M.J., Hennekens, C.H., Manson, J.E., Colditz, G.A., Rosner, B., & Willett, W.C. (1993). Vitamin E consumption and the risk of coronary disease in women. *New England Journal of Medicine*, **328**, 1444-1449.

Treatment of Mild Hypertension Research Group (1991). The Treatment of Mild Hypertension Study: A randomized, placebo-controlled trial of a nutritional-hygienic regimen along with various drug monotherapies. *Archives of Internal Medicine*, **151**, 1413-1423.

U.S. Department of Health and Human Services (1989). *Reducing the health consequences of smoking: 25 years of progress* (DHHS Publication No. CDC 89-8411). Washington, DC: U.S. Government Printing Office.

U.S. Department of Health and Human Services (1991). *Healthy people 2000: National health promotion and disease prevention objectives* (DHHS Publication No. PHS 91-50213). Washington, DC: U.S. Government Printing Office.

Willett, W.C., Hunter, D.J., Stampfer, M.J., Colditz, G., Manson, J.E., Spiegelman, D., Rosner, B., Hennekens, C.H., & Speizer, F.E. (1992). Dietary fat and fiber in relation to risk of breast cancer: An 8-year follow-up. *Journal of the American Medical Association*, **268**, 2037-2044.

Willett, W.C., Stampfer, M.J., Colditz, G.A., Rosner, B.A., & Speizer, F.E. (1990). Relation of meat, fat, and fiber intake to the risk of colon cancer in a prospective study among women. *New England Journal of Medicine*, **323**, 1664-1672.

Williamson, D.F., Madans, J., Anda, R.F., Kleinman, J.C., Kahn, H.S., & Byers, T. (1993). Recreational physical activity and ten-year weight change in a U.S. national cohort. *International Journal of Obesity*, **17**, 279-286.

Wood, P.D., Stefanick, M.L., Dreon, D.M., Frey-Hewitt, B., Garay, S.C., Williams, P.T., Superko, H.R., Fortmann, S.P., Albers, J.J., Vranizan,

K.M., Ellsworth, N.M., Terry, R.B., & Haskell, W.L. (1988). Changes in plasma lipids and lipoproteins in overweight men during weight loss through dieting as compared with exercise. *New England Journal of Medicine*, **319**, 1173-1179.

Wood, P.D., Stefanick, M.L., Williams, P.T., & Haskell, W.L. (1991). The effects of plasma lipoproteins of a prudent weight-reducing diet, with or without exercise, in overweight men and women. *New England Journal of Medicine*, **325**, 461-466.

Chapter 4

The Impact of Health Promotion on Health Care Utilization

Wendy D. Lynch

The impact of health promotion on the utilization of health care services has not been well documented. Empirically, we know that risky behaviors lead to an increased probability of disease. Logically, as you reduce the incidence of disease, you also reduce the need for medical services. Yet the impact of health risk status or changes in health risk status on the utilization of health services is not well understood.

One reason that the relationship between health promotion and health care utilization is not clear is that health promotion has assumed a simple, direct relationship between risk reduction and utilization reduction. The presumed mechanism is that reduced risk leads to reduced illness, which results in lower utilization and costs. The flaws in this assumption contribute to the frustrating lack of consensus on the impact of risk reduction on medical utilization.

Another reason for our lack of understanding is that the primary focus of health promotion investigations has been on costs. The outcome variables most frequently involve the total cost of medical care for the employees rather than numbers or types of services. This approach has resulted both from the limited availability of diagnostic data and from biases that costs are the main outcome of interest. Consequently, findings addressing utilization specifically are often hidden among studies that address overall costs and include utilization figures only as an afterthought.

This paper will address what is known about the relationship between health risk status and health care utilization. It will focus heavily on health care utilization as a behavior and provide a review of the literature that describes the complexities of that behavior. It will also present a model that clarifies the many components of utilization. This paper will try to summarize the literature regarding risk reduction programs, but it will also include information from strategies often not considered under the heading of traditional health promotion. These areas include self-care, self-management of chronic disease, self-efficacy, and other nontraditional approaches. And, finally, this paper will provide a list of critical issues that can guide future research in this area.

Before taking up these topics, I should define precisely what I mean by the terms *utilization* and *health promotion*:

1. **Utilization**: My use of this term will refer to actual measures of doctors' visits or hospital days, or actual utilization figures from medical claims data or direct self-report of medical services utilization. This paper will also use patterns of utilization for some discussions (e.g., the percentage of high-cost users). Given that there are more than 100 possible indicators of utilization (Maurana, Eichhorn, & Lonnquist, 1981), this represents a simple definition of utilization.
2. **Health Promotion**: The definition of health promotion used here includes all of the interventions typically defined under the context of traditional health promotion, for example, cardiovascular risk reduction. It will also include other interventions that have been referenced in the health promotion literature but that are less often considered under the heading of health promotion. These include self-care education, self-efficacy training, and self-management of disease.

Understanding Health Services Utilization

Although unfamiliar to the health promotion literature, the issue of health care utilization has been examined and discussed in the health services literature for decades. Many theoretical models predicting the use of services have been proposed and tested, with minimal success. In spite of their usefulness as predictors, these models do provide an historical outline for describing the complexity of utilization behavior. These studies also provided early warning that medical utilization is not simply a reflection of illness or health status.

Initial Investigations Revealing the Complexity of Utilization

In 1973, Andersen and Newman outlined a theoretical framework for understanding health services utilization. They hypothesized that utilization could be viewed as an individual behavior. The individual determinants of behavior are influenced by societal determinants, both directly and through the health care system. But the ultimate decision to seek care is an individual one. Andersen and Newman stated that use of services

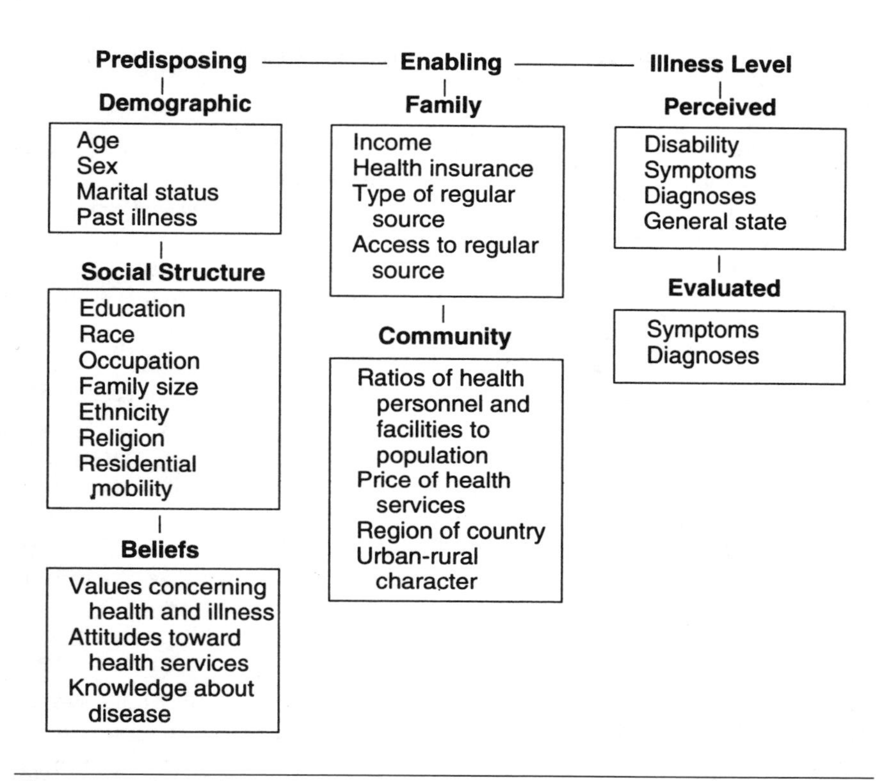

Figure 4.1 An early model explaining utilization behavior.
Note. Adapted from "Societal and Individual Determinants of Medical Care Utilization in the United States" by R. Andersen and J.F. Newman, 1973, *The Milbank Quarterly*, **51**, pp. 95-121. Copyright 1973 by The Milbank Memorial Fund. Adapted by permission.

is determined by the predisposition to use services (predisposing factors), the ability to secure services (enabling factors), and the level of illness (need). They further defined the components within each of the three determinants (see Figure 4.1), rated the importance of each, and provided a framework for further research.

This framework for defining utilization, although helpful in some regards, left many areas of ambiguity. Under the category of enabling factors, for instance, some factors relate to the medical care system—for example, the availability of physicians in the community or the size of one's deductible or copayment (Hulka & Wheat, 1985). Other enabling factors relate to one's family (e.g., socioeconomic status).

The concept of predisposing factors was also confusing because of the difficulty defining the actual source of the predisposition. For example, those who have been high users of health care are more likely to be high users in the future (McCall & Wai, 1983), and women tend to have more physician visits than men (Hulka & Wheat, 1985). While these are interesting observations, they provide no help in our understanding of why these specific individuals have the predisposition for more care.

As a factor, the need for medical services received by far the most attention in the literature. In some respects, the need for services is the only plausible causal predictor of utilization (Hulka & Wheat, 1985). The presence of symptoms increases the likelihood of seeking medical care. Also, a greater number of symptoms or greater severity of illness leads to increased utilization (Hulka & Wheat, 1985; Tanner, Cockerham, & Spaeth, 1983). In a study of randomly selected adults in Illinois, respondents recalled having one additional visit to a physician for each symptom reported (Tanner, Cockerham, & Spaeth, 1983). Other studies have found similar relationships between need and utilization (Berkanovic, Telesky, & Reeder, 1981). Despite being an obvious predictor, however, symptoms alone predict only a small portion of the variability in utilization (Tanner, Cockerham, & Spaeth, 1983). Given a particular symptom, it is difficult to predict whether an individual will seek care. The need for care also predicts some types of care more than others. Medical need, as one might expect, does not predict use of preventive services. Although disconcerting, need does not seem to predict surgical procedures either (Hulka & Wheat, 1985).

The framework provided by these early studies left many unanswered questions. Clearly, considerations other than medical need affect the decision to seek care and to use particular procedures. Up to this point, however, the nature of these other factors remained unclear. Methodologically, the need for services had been poorly defined. Such indicators as self-reported symptoms did not take into account possible differences in individuals' perceptions about their health status. These limitations led to a new direction in utilization investigations.

Investigations That Examined Utilization as a Behavior

Several large studies in the 1970s, both in the health services and the economic literature, examined the relationships between utilization and individual, societal, and organizational variables (Andersen, Kravits, & Anderson, 1975; Wolinsky, 1978). During the same period, smaller, more qualitative studies examined sociocultural and psychological predictors of illness behaviors (Robinson, 1971; Levine & Kozloff, 1978). Results from the two areas led to conflicting evidence about the contribution of individual perceptions and other psychosocial variables on care-seeking decisions.

In 1979, Mechanic strongly criticized the disjointed effort between the two areas of literature and described the disappointing lack of understanding of the processes and circumstances that influence decisions to seek care. Basically, he observed that studies large enough to have a representative national sample were unable to capture many of the important psychosocial influences on decision-making at the necessary level of detail or track changes in those variables over time.

Mechanic noted that results from the large studies of enabling factors (e.g., family income, insurance, M.D.-to-population ratio, etc.) had accounted for only 2% to 4% of the variability in physician utilization. Indeed, the most comprehensive models could account for only 16% to 25% of the variability in utilization, and most of the variance was explained by the diagnostic severity and individual's rating of illness. Mechanic also questioned the use of self-reported symptoms, chronic illness, or well-being as indicators of health status or need. His findings suggested that these self-reported indicators were inherently tied to perceptions and represented a poor proxy measure for true morbidity.

Mechanic called for a more behavioral approach that examined "the ways in which people perceive their bodies, make sense of their symptoms, and come to depend on the medical care system" (1979, p. 394). This statement reflects researchers' realizations that utilization behavior represents a complex decision-making process, one that is not easily explained by a small set of predictors.

Before reviewing subsequent literature in this area, I will present a model that defines the sources of variation in health care utilization. This model has been described in detail elsewhere and will provide a foundation on which further discussions will become clearer.

A Model Defining the Components of Utilization

It is difficult to discuss the impact of health promotion on utilization without providing a context for discussing utilization. Utilization is not a simple reflection of the need for services. However, the simplest explanation for utilization follows the principle of supply and demand: Anyone who is in need of service has access to service, and in reverse, access to service is available to anyone who needs service (see Figure 4.2). As such, reductions in utilization result from reductions in risks or from limitations on access.

This is a simple representation of utilization that health promotion has accepted for some time. However, the demand for medical services is not as simple as the existence of a medical condition or of morbidity. The supply of services is not as simple as access. In truth, the complex decision about whether to utilize services, as we have seen, includes everything from choice and knowledge and health status to biases and preference and financial incentives.

The traditional health promotion model relies on a simplified model of supply and demand. As you reduce someone's health risks, you improve his or her health status. As you improve health status, morbidity

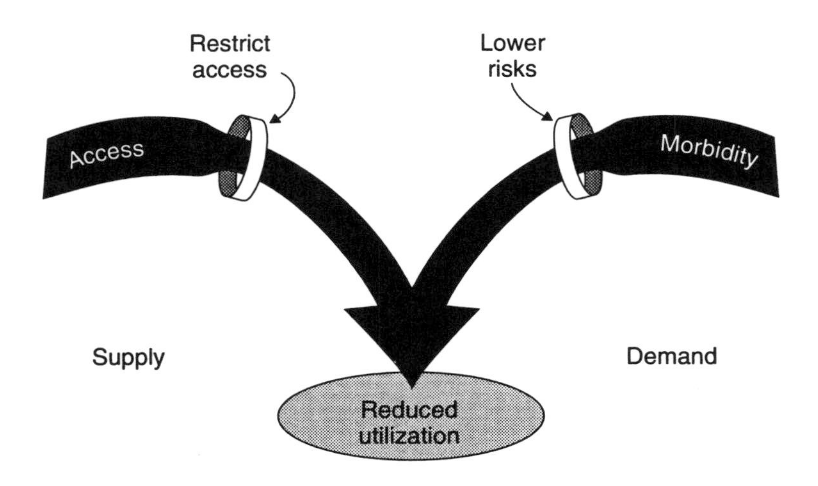

Figure 4.2 A simple representation of the drivers of health care utilization. The rationale behind traditional health promotion has often been based on the presumption that lower risks will lead to lower utilization.
Note. Adapted from "The Potential Impact of Health Promotion on Health Care Utilization: An Introduction to Demand" by W.D. Lynch and D.M. Vickery, 1993, *American Journal of Health Promotion*, **8**(2), p. 87. Copyright 1993 by *American Journal of Health Promotion*. Adapted by permission.

decreases. Unfortunately, the decrease in morbidity often does not translate into a decrease in utilization or a decrease in costs, leaving us to wonder what happened to our logic.

One key to understanding this problem lies in understanding the complex nature of utilization. As we have seen, decades of health services research have documented that utilization reflects a behavior by an individual. That behavior reflects several dimensions of demand. Demand for services is only partly influenced by the need for services. Figure 4.3 provides a functional model for understanding utilization. Supply-side components include many of the aspects of the enabling and predisposing factors defined by Andersen and Newman in 1973. Although supply-side factors account for a large portion of the variability in utilization, only the demand components will be discussed here. These include an objective indicator of the need for services, morbidity, as well as three other areas: the perceived need for services, patient preferences, and moral hazard.

Perceived need for services includes all personal, cultural, and social influences that shape an individual's belief that services should be sought. Patient preferences include the individual's choices and judgments regarding whether the benefits of a particular course of action outweigh

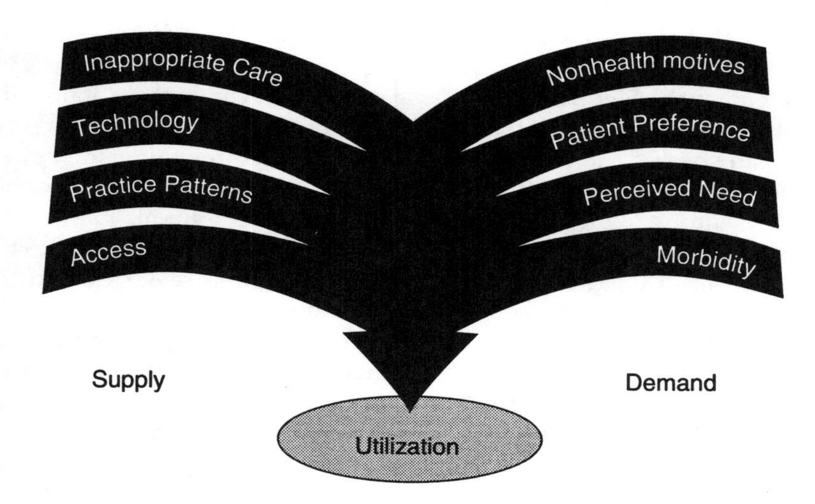

Figure 4.3 A representation of the many components that influence health care utilization behavior.
Note. Adapted from "The Potential Impact of Health Promotion on Health Care Utilization: An Introduction to Demand" by W.D. Lynch and D.M. Vickery, 1993, *American Journal of Health Promotion*, **8**(2), p. 89. Copyright 1993 by *American Journal of Health Promotion*. Adapted by permission.

the risks and costs of treatment. Nonhealth motives refer to situations where behaviors are influenced by attractive incentives and benefits available at little or no cost. Employees who pay very little for health care may not consider the cost ramifications of a medical choice. Further, they may choose to use services for reasons other than health (e.g., to get time off). A more detailed description of these components of demand have been presented elsewhere (Vickery & Lynch, in press). Here, we will use this model to review other studies and examine which components of demand were addressed.

Investigations Addressing Demand Components of Utilization

More recent studies of utilization behavior attempted to include a more comprehensive set of behavioral variables, most of which fall under the heading of perceived need. Berkanovic, Telesky, and Reeder (1981) examined a comprehensive set of behavioral, attitudinal, social, and organizational variables over a 1-year period to try and explain the decision to seek care. Overall, this model could explain 57% of the variability in the decision to seek medical care. Twelve percent of the variability could be attributed to morbidity variables (number of conditions, degree of disability).

Forty-two percent of the variability in utilization could be explained by perceived need—for example, influence from the individual's social network (a loved one gives advice to seek care) or personal beliefs, especially efficacy (belief that a physician can help). The authors found the results surprising because the influence of the network and beliefs were symptom-specific. An individual's general health orientation and beliefs did not contribute significantly to the model. Evidently, one's usual orientation toward care may not apply to any specific condition.

A variety of studies have demonstrated that other factors related to the perceived need for care contribute to the decision to seek care and the extent to which services are used. For example, one study found that perceived seriousness (termed *evaluated necessity*) combined with the presence of the symptom increased the likelihood of seeking care and increased the number of doctor visits by 20% (Tanner, Cockerham, & Spaeth, 1983).

Gender differences in utilization may also relate to differences in perceived need. Women generally seek more care than men, leaving questions about whether this reflects a difference in need or a difference in tendency to seek care. One hypothesized reason for gender differences in utilization behavior is that women perceive symptoms more seriously yet are less reluctant to act upon them (Hibbard & Pope, 1986). However, other evidence suggests that men are more likely to overrate the severity of their symptoms in comparison to clinical ratings of their minor symptoms and that the gender difference may not be simply a difference in perception (Macintyre, 1993).

In older persons, utilization may reflect a complex relationship among several indicators of both morbidity and perceived need. In a recent panel study, Counte and Glandon (1991) demonstrated that changes in health status explain some of the changes in utilization, especially hospital care. But, health status was not the strongest predictor of physician or emergency room visits. In their study, the occurrence of stressful life events was not predictive of physician visits as a single predictor. However, a combination of stressful life events and low social support (interaction), was the strongest predictor of utilization. The use of the medical system for psychosocial support has been hypothesized and suggested in several other studies (Kurz, Haddock, Van Winkle, & Wang, 1991).

Another dimension of utilization behavior is the degree of medical service use following diagnosis. Although one might assume that use of services for serious chronic conditions reflects the severity of the illness, this is not clear. In a prospective study of 215 seriously ill patients, Browne et al. (1990) discovered that physician and hospital services were unrelated to severity or prognosis and most strongly associated with the patient's adjustment to illness, measured using an instrument that examines such attributes as the meaning the patient attaches to his or her condition. Although the lack of association with severity could reflect inadequate measurement, adjustment to illness accounted for a wide difference in utilization. Poorly adjusted patients consumed six times as many specialist visits and nine times as many visits for other health professionals in a 6-month period than well-adjusted patients. Clearly, this was not a difference in morbidity but a difference in the perception that additional care was needed.

In many instances the type and amount of treatment can vary widely following a medical diagnosis. Obviously, treatments can be influenced considerably by physician advice, insurance coverage, and other supply-side factors. But a portion of that decision is driven by a patient's decision concerning the personal value of the treatment. Studies indicate that patients are more risk averse than physicians. Patients given complete information concerning risks and benefits choose to postpone or avoid more invasive treatments significantly more often than patients not receiving that information (Denzer, 1991). In this case, utilization behavior reflects patient preference rather than morbidity or perceived need.

Need Versus Demand Versus Utilization

Two important distinctions derived from the health utilization literature that deserve reiteration are the clear difference between need for services and demand for services, and the difference between demand for services and actual utilization. As pointed out by Mechanic 14 years ago, reported presence and severity of symptoms are inherently subjective and usually differ from the "objective" ratings by clinicians. In this chapter, we will use morbidity to represent need. Morbidity, in turn, represents one component of the demand for services. Demand for services represents only half of the dynamic that results in health care utilization. The other half of the dynamic involves components that influence utilization through the supply of health care services (see Figure 4.3).

Traditional health promotion, as mentioned previously, focuses primarily on reducing the need for services. Viewed in this way and acknowledging the complexity of utilization, it is obvious that health promotion can only address demand through risk reduction and direct prevention of morbidity. In this context, risk reduction has only limited potential to affect the entire range of health care utilization and the dynamics of utilization behavior.

Interventions that have an impact on the several components of utilization, such as perceived need or patient preference, obviously will have an impact on a wider spectrum of utilization behaviors and perhaps a greater portion of medical care usage.

The Impact of Health Promotion on Health Services Utilization

This section of the chapter includes a brief discussion of the logical premise behind this body of literature and the measurement issues that affect its quality. This section will also review the studies that have examined the relationship between health promotion and health care utilization, categorized by their measurement strategy.

Logical Basis

The premise underlying health promotion is that certain behaviors and exposures minimize the likelihood of disease. Thousands of studies, spanning from the effects of smoking to the risk of injury in competitive sports, support the notion that the probability of morbidity can be reduced or eliminated based on lifestyle and behavioral choices.

Perhaps the most comprehensive review of this literature is the Carter Center's document, *Closing the Gap*. This compilation of papers summarizes the knowledge to date concerning risk factors that have a scientifically documented association with one or more diseases. For cancer and cardiovascular disease, a total of seven modifiable risk factors account for 23% and 65% of morbidity, respectively (Amler & Dull, 1987). It follows that interventions that reduce the exposure or prevalence of the risky behaviors would also reduce the likelihood of their associated medical outcomes.

The additional two steps in logic that health promotion as a discipline has taken, especially in the corporate sector, is to assert that interventions that reduce risky behaviors and lead to reductions in illness and injury will also reduce the need for medical services and result in lower heath care costs (Lynch, 1993). The first part of this argument is difficult to dispute given the strength of medical evidence. We know, for instance, that a successful reduction in cardiovascular risks for a population of employees will result in fewer heart attacks than would have otherwise occurred. It is reasonable to assume that the need for cardiovascular-specific health services would be reduced.

The second part of the argument is less certain. The degree to which changes in the medical need for services will be reflected in overall utilization is questionable. If actual health status, based on symptoms, accounts for only 12% of the variability in utilization at most (Berkanovic, Telesky, & Reeder, 1981), it is possible that even true avoidance of disease would not produce dramatic changes in utilization. The issue is not whether the health promotion program improved health, nor whether morbidity was reduced. Instead the issue is whether overall utilization, which is a complex behavior that only partially reflects actual morbidity, would change as a result of the avoidance of a specific health condition.

Methodological Considerations

Table 4.1 outlines the continuum of measurement strategies for detecting the impact of risk-reduction programs on utilization. The table does not address all aspects of measurement but focuses on two areas of greatest concern for the studies reviewed here. The choice of outcome measure is of particular concern. Obviously, precise, condition-specific utilization indicators would have the best chance of detecting the impact of health risk interventions, especially if these outcome measures were linked to actual risk-reduction at the individual level and the outcomes followed a time period sufficient to expect a true reduction in morbidity. Unfortunately, such studies have not yet been published.

As shown in Table 4.1, methodologies that employ less-specific outcome measures or less-specific indicators of risk add inherent weakness to the validity of conclusions. Weaker measurement strategies include studies that did not have risk-specific utilization data and had self-reported predictors or outcomes. When reviewing studies that have a variety of limitations, it becomes difficult to rank them from "best" to "worst." Each provides different information and has different problems. For the purposes of this review, studies will be categorized by their utilization outcome measurement strategy: any outcome including risk-specific utilization, any outcome of overall utilization or cost outcome, or high-cost utilization categories.

Table 4.1
Research on the Relationship Between Health Promotion Health Care Utilization
Measurement Considerations

Rating scale	Risk predictors	Utilization outcomes
Strongest	Changes in actual risk (pre and post)	Changes in actual risk-specific utilization (pre and post)
	Preintervention actual risk	Postintervention longitudinal actual risk-specific utilization
	Preintervention self-reported risk	Postintervention actual risk-specific costs
	Participation in risk reduction	Self-reported utilization
	Self-reported participation in risk reduction	High-cost utilization categories
		Overall costs
Weakest		Self-reported overall costs

Studies Having a Risk-Specific Utilization Outcome

As mentioned, no studies have used ideal measures of the impact of health promotion on utilization. The studies in this section used condition-specific outcomes but did not have individual risk-reduction data. Clearly, it is difficult to infer causality when the cause is not measured directly.

The risk-specific approach was used by prevention-related claims analyses in a worksite setting (Sciacca, Seehafer, Reed, & Mulvaney, 1993). This method isolates the conditions that have a reasonably certain connection to the intervention and quantifies the impact with an appropriate outcome measure reflecting a reduction in morbidity. Sciacca and colleagues measured costs for preventable illness over 5 years but linked the information to participation rather than risk level. They found no reduction in preventable costs for participants. Johnson and Johnson's lifestyle claims analysis quantifies the portion of claims that are related to preventable conditions, using incidence data; however, reports to date have not been linked to individual risks or monitored change over time. No reports on the risk-specific approach have been published (Peterson, personal communication).

Studies Having Overall Utilization Outcomes

Because the outcome measure lacks specificity, these studies have major threats to the validity of their findings. They include studies that did not focus on risk-specific utilization but did measure individual risks and a subsequent indicator of overall medical utilization. Utilization might be measured as overall visits, hospital days, or self-reported use. The inherent weakness in this type of outcome measure is the assumption that risk-reduction will reduce utilization overall.

The ongoing study of a sample of Bank of America retirees has provided some information about health risks and subsequent self-reported utilization. Both the 12-month and 2-year results of a health promotion and self-care intervention indicated a significant decrease in the number of reported sick days (days with restricted activity); however, observed reductions in physician visits and hospital days were not significant (Leigh & Fries, 1992).

Of the individual risk factors, only body mass was significantly associated with physician visits, and both body mass and cigarette smoking were associated with increased hospital days. As noted by the authors, all relationships between risk factors and utilization, though not statistically significant, were in the expected direction (Leigh & Fries, 1992).

Two-year findings from this study indicate that age, gender, education, and five risk factors (body mass, exercise, smoking, seat belts, and alcohol consumption) account for only 2% of the variation in doctor visits, 16% of the variation in hospital days, and 2% of the variation in direct medical costs (Leigh & Fries, personal inquiry, 1992). The same variables predict 25% of the variation in sick days. This suggests that health risks contribute more to one's ability to function (sick days) than to the decision to seek care (actual utilization).

Studies of blood pressure and utilization also provide interesting results. Erfurt and Foote (1989) documented that hypertensives under medical treatment visited the doctor more frequently overall than normotensives. However, they also visited the doctor more frequently than hypertensives not under treatment. This finding, although not intuitively obvious, documents the phenomenon that screening sometimes increases utilization by identifying the morbidity (Erfurt & Foote, 1989). Furthermore, the number of visits was the same for individuals whose hypertension was under control as those whose hypertension was not controlled. One might suggest that in the short term, utilization was driven solely by the presence of the treatment and not by the presence of the disease. In the long term, however, evidence does suggest that controlled hypertension may reduce the incidence of cardiovascular hospitalization (Alderman, Madhavan, & Davis, 1983).

Pope (1982) examined the relationship between three lifestyle habits (smoking, alcohol consumption, and physical activity) and utilization of health service for 2,500 members of an HMO. No consistent relationships were found between the lifestyle indicators and eight categories of utilization across age groups or gender. A few weak associations were reported between specific risks and categories of use ($r < .20$).

Results from the 1977 National Health Interview Survey found that those with the most healthy behaviors tended toward fewer doctor visits and fewer hospital days (Wetzler & Cruess, 1985). Physical activity (moderate to high) was related to fewer doctor visits, and hours of sleep was associated with reduced doctor visits and hospital days. Body weight and cigarette smoking were not associated with utilization.

In a recent study comparing the utilization patterns of employees who responded to a health risk appraisal versus employees who did not respond, the authors found that responders had an increased likelihood of seeking any care and greater use of low-cost services than nonresponders despite lower overall levels of risk (Lynch et al., 1993). Further investigation revealed that responders were also more likely to seek care for the most common diagnoses, such as sore throat, respiratory infection, abdominal pain, etc. (Lynch, 1993). These observations were surprising given the risk levels of the two groups, but they probably reflect determinants of utilization other than risk status (Lynch, 1993).

Studies Having a High-Cost Utilization Outcome

The health promotion literature has examined utilization from the perspective of high-cost utilization versus low-cost utilization. Although the definitions of high cost have varied, essentially these studies have identified a small portion of individuals who use a disproportionate amount of medical care. The idea behind this approach is to identify whether those with health risks are more likely to have exceptionally serious, expensive medical outcomes (Lynch, Main, & Teitelbaum, 1992).

A small number of individuals account for a large percentage of health care costs. Estimates of the distribution of costs vary, with one study finding that 18% of individuals accounted for 88% of costs (McCall & Wai, 1983), and another tracing 64% of costs to 10% of the people (Yen, Edington, & Witting, 1991). In a 6-year study of the elderly, 26% of the population were consistent high-cost users (4 of 6 years) and this group accounted for 50% of costs.

The association of smoking with high-cost cases has been examined in several studies. Three separate studies have documented that smokers are at greater risk of high cost than nonsmokers (Lynch et al., 1992; Yen et al., 1991; Freeborn et al., 1990b). This relationship holds true for employee populations (Lynch et al., 1992; Yen et al., 1991) and for those over 65 years of age (Freeborn et al., 1990a). These studies also included subjects from both group insurance and group practice HMOs. Such information supports estimates that smokers have 40% greater lifetime medical expenditures than nonsmokers (Hodgson, 1992).

Yen and colleagues (1991) found that alcohol consumption also produces a significant increase in the likelihood of high costs. They examined the costs of other health risks but the increased likelihood of high cost was not statistically significant. Interestingly, aside from smoking, the strongest predictors of high costs were attitudinal: low job satisfaction, perceptions of poor health, low life satisfaction, and high stress.

Another study of high-cost users documented that high users had more chronic conditions but also perceived themselves in poorer health and in greater psychological distress than low utilizers (Freeborn et al., 1990b). However, because perceptions were measured after utilization occurred, it is difficult to know whether the utilization resulted from the distress or vice-versa.

Using high-cost utilization as an outcome measure makes sense when the expected medical consequences are serious or catastrophic in nature (Lynch et al., 1992). It is a safe generalization that treatments for most of the severe risk-related morbid conditions (e.g., stroke, lung cancer, premature birth) result in high medical expenses. In the case of smoking, the expected medical outcomes are usually costly, and studies consistently confirm an increased likelihood of high costs. Health risks with less serious or unknown outcomes may not be appropriately assessed with this approach. Once again, indicators related to need (risks) were not the most reliable predictors of high-cost utilization. Attitudes, which may contribute to perceived need, consistently appeared as predictors of high-cost use.

Summary of Risk-Reduction Studies

From the limited number of studies in this area and the severity of their limitations, it is difficult to draw solid conclusions about the impact of risk-reduction on utilization, especially in the short term. Although we know that risk-reduction has some impact on morbidity, risk levels themselves may have little association with overall utilization behaviors. Thus, the measurement strategies used by the majority of studies may not have had the capability of detecting any impact of risk-reduction on utilization.

Utilization-Specific Interventions

Although more precise and certainly more sensitive to risk-specific utilization, focusing on risk-defined medical outcomes leaves out most of health care utilization. The true impact of health promotion could reflect a much larger portion of utilization—portions that reflect perceived need for services, perceived efficacy of treatments, self-care skills, and more. All of these aspects of demand will be reviewed in what follows.

Based on the arguments presented thus far, readers will recognize that interventions designed specifically to improve health will not necessarily reduce utilization. To decrease utilization, successful interventions may need to target utilization behaviors more directly. Interventions that target utilization, unlike the traditional interventions, may not be thought of as cost-containment interventions but as true health-promoting efforts (Warner, 1993). Nonetheless, it can be argued that these interventions do have a positive impact on health.

Encouraging appropriate use of medical care and reducing unnecessary encounters with the medical system is in the best interests of the individual in many respects. There are risks inherent to medical care to which one is exposed (Gofman & O'Connor, 1985; Lambert, 1978; Vickery, 1990). In many cases the potential benefits of treatment far outweigh the risks. Nevertheless, unnecessary contact with the medical system leads to unnecessary exposure to those risks.

Although debatable, there are reasons to consider interventions that encourage appropriate use of medical care as health-promoting interventions. Not the least of these are the well-being and empowerment of an individual who learns to manage a chronic condition or make an informed decision concerning treatment.

For these reasons, the following sections have been included as health promotion strategies rather than pure cost-containment tactics. Under the context of demand management, rather than need reduction, the following interventions have the greatest potential for affecting the perceived need for services.

Self-Care

Studies of health care utilization focus on individuals who have made contact with the medical care system. Yet most clinical symptoms are actually treated at home. Health surveys indicate that on any given day, nine out of ten supposedly healthy individuals have at least one clinical symptom (Kart, Metress, & Metress, 1978), and that only 10% to 25% of symptoms result in professional health care (DeFriese & Woomert, 1983).

Self-care responses to illness include bed rest, home remedies, prayer, use of over-the-counter medication, use of prescription medication from a friend or family member, or no action at all (Stoller, Forster, & Portugal, 1993). There is marked variation in self-treatment strategies among individuals and considerable variation in strategies for the same individual over time (Stoller et al., 1993). Decisions to self-treat a symptom are influenced by factors similar to those in the decision to seek care: perceived seriousness of the condition, prior experience in treatment, and level of disability or discomfort caused by the symptom (Stoller et al., 1993).

Interestingly, in a study of the elderly, fewer than one sixth of the subjects reported managing symptoms through lifestyle or self-education; "they were more likely to manage symptoms with medication than to prevent their occurrence in the future" (Stoller et al., 1993, p. 37). The tendency to self-medicate has elicited valid concerns about the potential risks associated with unsupervised use of prescription drugs (Haug, Wykle, & Namazi, 1989). However, as a whole, evidence suggests that self-care and self-treatment decisions are generally appropriate and beneficial (Dean, 1986; Ford, 1986).

In this context, programs and resources that provide education concerning self-care choices offer valuable guidance for decisions individuals make about symptoms virtually every day. Medical self-care programs refer to materials and resources that promote appropriate decision-making, both about the choice to seek professional care and the alternatives for self-management of symptoms.

Self-care reference books have been successful tools for reducing physician visits. In an adult working population, Vickery et al. (1983) reported a 17% reduction in physician visits overall and a 35% reduction in visits for minor conditions. Similar findings have been reported in a Medicare population (Vickery et al., 1988). Although direct measures of quality were not used, these studies suggest that the reduction in physician care had no detrimental effect on health status.

Self-Management and Self-Efficacy

In the latter part of this century, acute and infectious diseases have been replaced by chronic diseases as the number-one cause of disability in this country (Vickery, 1990) and the major reason for use of medical care by the elderly. Consequently, reducing treatment for chronic illness has great potential for reducing overall health care utilization. As previously discussed, the ideal method for reducing the need for chronic illness care is by preventing the illness altogether through behavior change. As a long-term strategy this has much appeal, but it overlooks the cases of illness that already exist.

A more direct and immediate approach for reducing utilization for chronic illness is to promote self-management of symptoms. For example, self-monitoring of blood sugar or blood pressure increases the diabetic's or hypertensive's involvement in their condition and may lessen the demand for ongoing supervision by medical professionals. In such cases individuals assume some responsibility for monitoring their condition and may learn to recognize and prevent certain events related to their disease (e.g., how to respond to especially low blood sugar) (Greenfield et al., 1988). At best, the individual can also choose appropriate medical care based on known warning signals.

Self-management of disease has been especially beneficial for conditions that are not life-threatening but which present tremendous discomfort or disability, such as chronic pain and arthritis. These conditions interfere with daily activity, worsen over time, and show limited response to medical intervention. Thus, treatments are limited to symptom management or, in the case of arthritis, extreme interventions such as total replacement of the natural joint.

Research in self-management of arthritis has demonstrated that educational interventions can reduce physician visits quite dramatically. Holman and colleagues have documented a 43% decrease in physician visits over 4 years compared to arthritics who did not participate in a self-management course (1989). The patients also reported significant decreases in pain and slower progression of disability.

Perhaps the most important finding in this research was the strong role that self-efficacy played in patients' improved health status and reduced demand for medical care. Surprisingly, changes in health status were not strongly associated with changes in health behaviors (Lorig et al., 1989). The process by which patients improved health status was through improvements in self-efficacy—changes in their level of confidence that they could manage their own symptoms. Subsequent studies confirmed that specific efficacy-enhancing strategies increased the effectiveness of the intervention (Lorig & Gonzalez, 1992).

Self-efficacy refers to an individual's confidence in his or her ability to perform specific behaviors (Bandura, 1989). These beliefs can be influenced by mastery experience (practicing until you know you can do it), modeling (seeing that someone else can do it), social persuasion (hearing someone else say you can do it), and psychological state (being in the right mindset to do it). Self-efficacy is skill-specific. Confidence in the ability to lose weight does not necessarily translate into confidence in the ability to quit smoking.

Since it was postulated by Bandura in 1977, self-efficacy theory has demonstrated many relevant applications to health care utilization. Perceived self-efficacy has been shown to influence both someone's willingness to try to change a health behavior as well as the likelihood of making and maintaining that change (Bandura, 1989). Studies show that perceived self-efficacy plays a role in many different health settings, such as the rate of recovery from myocardial infarction (Debusk, Kraemer, & Nash, 1983), the ability to cope with pain with less medication (Manning & Wright, 1983), and the levels of distress in potentially adverse situations (Bandura, Reese, & Adams, 1982).

Self-efficacy also alters health responses at a more basic biological level. Self-doubts in coping abilities actually produce instantaneous changes in blood chemistry. Wiedenfeld et al. (1990) documented dramatic changes in the components of the immune system as a result of coping self-efficacy training.

These examples emphasize the importance of self-efficacy in illness behaviors. Although more frequently discussed as a component of health education and behavior change interventions, self-efficacy training has tremendous potential for improving health status and reducing health services utilization. In essence, self-efficacy seems to play a role both in helping the individual reduce both the true need for services and the perceived need for services.

Areas for Future Research

To obtain a better understanding of utilization, researchers should employ direct indicators of use of services rather than summary costs. Total costs give little information about the frequency of utilization or the nature of the condition or of the care received. Health promotion's focus on costs thus far has in part reflected the difficulty in obtaining detailed cost information. The increased availability of detailed information should lead to more detailed analyses of utilization patterns and characteristics.

If the health promotion discipline wishes to establish its economic value based on its impact on medical costs and utilization, research efforts must recognize the complexity of utilization behavior. Health promotion has not drawn upon the existing body of knowledge concerning utilization behavior, so it cannot yet define its potential impact on that behavior. Evaluations should distinguish between the concepts of reducing morbidity and reducing demand and recognize that morbidity represents only one small component of utilization. Figure 4.4 depicts some of the possible influences that health promotion can have on utilization. By separating utilization and demand into these functional components, researchers can

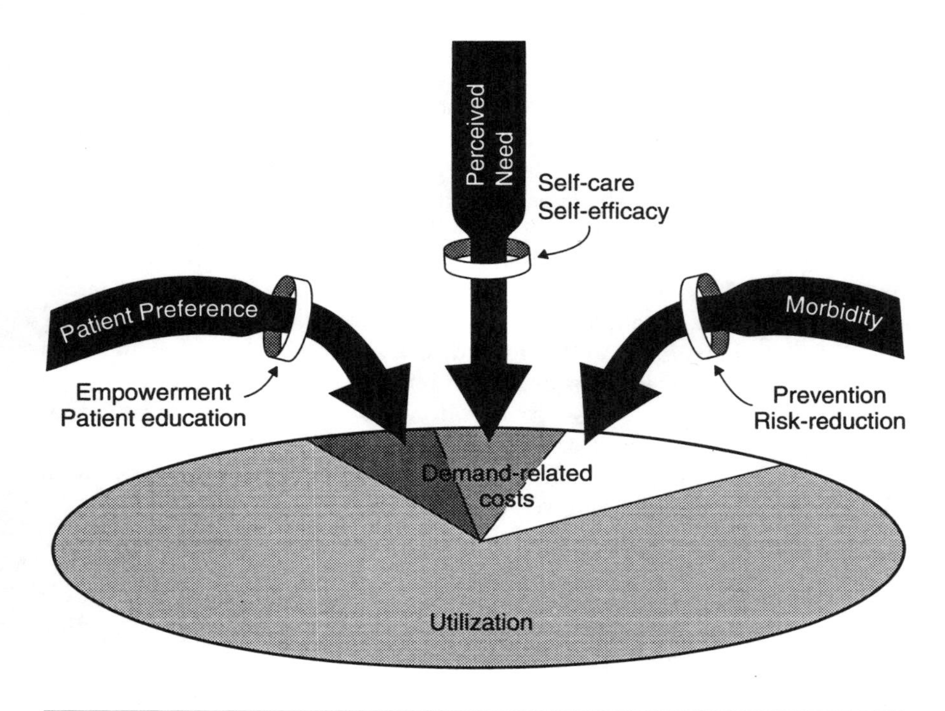

Figure 4.4 Measuring the impact of health promotion on utilization.

begin to define more specific utilization outcome measures. These provide opportunities to more accurately define and assess health promotion's impact on utilization.

Specifically, studies should define the role of health promotion in managing the demand for medical services. This should focus on the following questions:

1. How did utilization change? Did the proportion of individuals who sought care change, or was the frequency of visits reduced?
2. What component of demand has been affected? Was there a true reduction in morbidity (e.g., need for services), or was there a decrease in the perceived need for professional care? Did patient preference or nonhealth motives play a role?
3. What mechanism was used to invoke this change in utilization? Was there a reduction in a risk behavior, or was there an increase in confidence that the symptom could be self-managed?

Each of these issues serves to identify more precisely the processes by which health promotion influences illness behavior and medical care utilization.

Summary

The key to understanding the relationship between health promotion and health care utilization lies in understanding utilization behavior. Effective risk-reduction programs have the potential for reducing the likelihood of serious and expensive diseases, which in turn can impact the need for services. Based on decades of research concerning health care utilization, the level of morbidity influences actual utilization much less than one might expect. Thus, the measurable impact of health promotion on overall utilization is also lessened.

Utilization behaviors can be more directly influenced by altering one's knowledge, skills, resources, and confidence in the ability to self-manage a symptom or disease. These interventions do not fall under a traditional heading of health promotion but have the potential for improving perceived health status and disability and reducing unnecessary contact with the medical care system. Figure 4.4 illustrates how health promotion may affect utilization through several different mechanisms, not only through disease prevention.

In light of the complexity of utilization behavior, the inconsistency among results of studies that examine the relationship between health promotion and costs should come as no surprise. Similarly, the lack of evidence concerning health promotion and utilization might be expected. Utilization of health services is not the predictable, inevitable result of a given disease. Utilization represents a behavior that results from a dynamic and personal decision-making process, influenced by a myriad of other considerations, circumstances, and beliefs. Clearly, the measurable components of demand for services must be more clearly defined by researchers in order to assess the true impact of health promotion.

References

Alderman, M., Madhaven, S., & Davis, T. (1983). Reduction of cardiovascular disease events by worksite hypertension treatment. *Hypertension*, **5**(Suppl. V), V138-V143.

Amler, R.W., & Dull, H.B. (Eds.) (1987). *Closing the gap: The burden of unnecessary illness*. New York: Oxford University Press..

Andersen, R., Kravits, J., & Anderson, O. (Eds.) (1975). *Equity in health services: Empirical analyses in social policy*. Cambridge: Balinger.

Andersen, R., & Newman, J.F. (1973). Societal and individual determinants of medical care utilization in the United States. *The Milbank Quarterly*, **51**, 95-121.

Bandura, A. (1977). Self-efficacy: Toward a unifying theory of behavior change. *Psychology Review*, **84**, 191-215.

Bandura, A. (1982). Self-efficacy in human agency. *American Psychology*, **37**, 122-147.

Bandura, A. (1989). Self-efficacy mechanism in physiological activation and health-promotion behavior. In J. Madden, IV, S. Matthysse, & J. Barchas (Eds.) *Adaptation, learning and effect*. New York: Raven Press.

Bandura, A., Reese, L., & Adams, N.E. (1982). Microanalysis of action and fear arousal as a function of differential levels of perceived self-efficacy. *Journal of Personality and Social Psychology*, **43**, 5-21.

Berkanovic, E., Telesky, C., & Reeder, S. (1981). Structural and social psychological factors in the decision to seek medical care for symptoms. *Medical Care*, **19**, 693-709.

Browne, G.B., Arpin, K., Corey, P., Fitch, M., & Gafni, A. (1990). Individual correlates of health service utilization and the cost of poor adjustment to chronic illness. *Medical Care*, **28**, 43-58.

Bush, P.J., & Iannotti, R.J. (1990). A children's health belief model. *Medical Care*, **28**, 69-83.

Counte, M.A., & Glandon, G.L. (1991). A panel study of life stress, social support, and the health services utilization of older persons. *Medical Care*, **29**, 348-361.

Dean, K. (1986). Lay care in illness. *Social Science Medicine*, **22**, 275.

DeBusk, R.F., Kraemer, H.C., & Nash, E. (1983). Stepwise risk stratification soon after acute myocardial infarction. *The American Journal of Cardiology*, **12**, 1161-1166.

DeFriese, G., & Woomert, A. (1983). Self-care among the U.S. elderly. *Research on Aging*, **5**, 3.

Denzer, S. (1991). The allegory of the conveyor belt. *Dartmouth Alumni Magazine*, **5**, 23-25.

Erfurt, J.C., & Foote, A. (1989). *Long-term effects of hypertension programs in industry* (Final Report). Worker Health Program, Institute of Labor and Industrial Relations, The University of Michigan, Ann Arbor, MI. Grants from National Heart, Lung and Blood Institute, National Institutes of Health, U.S. Dept. of Health and Human Services.

Ford, G. (1986). Illness behavior in the elderly. In K. Dean, T. Hickey, & B. Holstein (Eds.), *Self-care and health in old age* (p. 130). London: Croom Helm.

Freeborn, D.K., Mullooly, J. P., Pope, C.R., & McFarland, B.H. (1990a). Smoking and consistently high use of medical care among older HMO members. *American Journal of Public Health*, **80**, 603-605.

Freeborn, D.K., Pope, C.R., Mullooly, J.P., & McFarland, B.H. (1990b). Consistently high users of medical care among the elderly. *Medical Care*, **28**, 527-540.

Fries, J.F. (1990, March). *The compression of morbidity: Progress and potential*. Paper adapted from address paper presented to the 1990 Sandoz Lectures in Gerontology, Basle, Switzerland, and to Compression of

Morbidity Conference, Asilomar, California. (This paper was supported in part by a grant from the National Institutes of Health [AM21395] to American Rheumatism and Aging Medical Information System.)

Fries, J.F., Williams, C.A., & Morfeld, D. (1992). Improvement in intergenerational health. *American Journal of Public Health*, **82**, 109-112.

Gofman, J.W., & O'Connor, E. (1985). *X-Rays: Health effects of common exams*. San Francisco, CA: Sierra Club Books.

Greenfield, S., Kaplan S.H., Ware, Jr., J.E., Yano, E.M., & Frank, H.J. (1988). Patient participation in medical care: Effects on blood sugar control and quality of life in diabetes. *Journal of General Internal Medicine*, **3**, 448-457.

Haug, M. (Ed.) (1981). *Elderly patients and their doctors*. New York: Springer.

Haug, M., Wykle, M., & Namazi, H. (1989). Self care among older adults. *Social Science Medicine*, **29**, 171.

Hayes, D., & Ross, C.E. (1986). Body and mind: The effect of exercise, overweight, and physical health on psychological well-being. *Journal of Health and Social Behavior*, **27**, 387-400.

Hibbard, J., & Pope, C. (1986). Another look at sex differences in the use of medical care: Illness orientation and the types of morbidities for which services are used. *Women & Health*, **11**(2), 21-37.

Hodgson, T.A. (1992). Cigarette smoking and lifetime medical expenditures. *The Milbank Quarterly*, **70**, 81-125.

Holman, H., Mazonson, P., & Lorig, K. (1989). Health education for self-management has significant early and sustained benefits in chronic arthritis. *Transactions of the Association of American Physicians*, **102**(204), 204-208.

Hulka, B.S., & Wheat, J.R. (1985). Patterns of utilization—the patient perspective. *Medical Care*, **23**, 438-459.

Jarvis, K.B., Phillips, R.B., & Morris, E.K. (1991). Cost per case comparison of back injury claims of chiropractic versus medical management for conditions with identical diagnostic codes. *Journal of Occupational Medicine*, **33**, 847-852.

Johnson, R.E., & Pope, C.R. (1983). Health status and social factors in nonprescribed drug use. *Medical Care*, **21**, 225-233.

Kart, C., Metress, E., & Metress, J. (1978). *Aging and health: Biological and social perspectives*. Boston: Addison-Wesley.

Kurz, R.S., Haddock, C., Van Winkle, D.L., & Wang, G. (1991). The effects of hearing impairment on health services utilization. *Medical Care*, **29**, 878-889.

Lambert, E. (1978). *Modern medical mistakes*. Bloomington, IN: Indiana University Press.

Leigh, J.P., & Fries, J.F. (1992). Health habits, health care use and costs in a sample of retirees. *Inquiry*, **29**, 44-54.

Leigh, J.P., Richardson, N., Beck, R., Kerr, C., Harrington, H., Parcell, C.L., & Fries, J.F. (1992). Randomized controlled study of a retiree health promotion program. *Archives of Internal Medicine*, **152**, 1201-1206.

Levine, S., & Kozloff, M.A. (1978). The sick role: Assessment and overview. *Annual Review of Sociology*, **4**, 317-343.

Lin, E.H.B., Katon, W., Von Korff, M., Bush, T., Lipscomb, P., Russo, J., & Wagner, E. (1991). Frustrating patients: Physician and patient perspectives among distressed high users of medical services. *Journal of General Internal Medicine*, **6**, 241-246.

Lorig, K., & Gonzalez, V. (1992). The integration of theory with practice: A twelve year case study. *Health Ed Q.*, **19**(3), 355-368.

Lorig, K., Selznick M., Lubeck, D., Ung, E., Chastain, R., & Holman, H.R. (1989). The beneficial outcomes of the arthritis self-management course are inadequately explained by behavior change. *Arthritis and Rheumatism*, **32**, 91-95.

Lynch, W.D. (1993). Using cost as a health promotion outcome: The problems with measuring health in dollars. In J. Opatz (Ed.), *Economic Impact of Worksite Health Promotion* (pp. 51-63). Champaign, IL: Human Kinetics.

Lynch, W.D., Gilfillian, L.A., Jennett, C., & McGloin, J. (1993). Health risks and health insurance claims costs. *Journal of Occupational Medicine*, **35**, 28-33.

Lynch, W.D., Main, D.S., & Teitelbaum, H.S. (1992). Comparing medical costs by analyzing high-cost cases. *American Journal of Health Promotion*, **6**, 206-213.

Macintyre, S. (1993). Gender differences in the perceptions of common cold symptoms. *Social Science Medicine*, **36**, 15-20.

Manning, M.M., & Wright, T.L. (1983). Self-efficacy expectancies, outcome expectancies, and the persistence of pain control in childbirth. *Journal of Personality and Social Psychology*, **45**, 421-431.

Maurana, C.A., Eichhorn, R.L., & Lonnquist L.E. (1981). The use of health services indices and correlates: A research bibliography. Springfield, VA: U.S. Dept. HSS, National Technical Information Service.

McCall, N., & Wai, H.S. (1983). An analysis of the use of medicare services by the continuously enrolled aged. *Medical Care*, **21**, 567-585.

Mechanic, D. (1979). Correlates of physician utilization: Why do major multivariate studies of physician utilization find trivial psychosocial and organizational effects? *Journal of Health and Social Behaviors*, **20**, 387-396.

Pope, C.R. (1982). Life-styles, health status and medical care utilization. *Medical Care*, **20**, 402-413.

Robinson, D. (1971). *The Process of Becoming Ill*. London: Routledge and Kegan Paul.

Rundell, T.G. (1979). A suggestion for improving the behavioral model of physician utilization. *Journal of Health and Social Behavior*, **20**, 103-104.
Sciacca, J., Seehafer, R., Reed, R., & Mulvaney, D. (1993). Health care costs and participation in a worksite health promotion program. *American Journal of Health Promotion*, **7**, 374-383.
Stoller, E.P., Forster, L.E., & Portugal, S. (1993). Self-care responses to symptoms by older people. *Medical Care*, **31**, 24-42.
Tanner, J.L., Cockerham, W.C., & Spaeth, J.L. (1983). Predicting physician utilization. *Medical Care*, **21**, 360-369.
Taylor, D.G., Aday, L.A., & Andersen, R. (1975). A social indicator of access to medical care. *Journal of Health and Social Behavior*, **16**, 39-49.
Vickery, D.M. (1990). *LifePlan*. Reston, VA: Vicktor, Inc.
Vickery, D.M., Golaszewski, T.J., Wright, E.C., & Kalmer, H. (1988). The effect of self-care interventions on the use of medical service within a medicare population. *Medical Care*, **26**, 580-588.
Vickery, D.M., Golaszewski, T.J., Wright, E.C., & Kalmer, H. (1989). A preliminary study on the timeliness of ambulatory care utilization following medical self-care interventions. *American Journal of Health Promotion*, **3**, 26-31.
Vickery, D.M., Kalmer, H., Lowry, D., Constantine, M., Wright, E., & Loren, W. (1983). Effect of a self-care education program on medical visits. *Journal of the American Medical Association*, **250**, 2952-2956.
Vickery, D.M., & Lynch, W.D. (Submitted 1994). Demand management and health care reform. *Annals of Internal Medicine*.
Warner, K. (1993). *Keynote address. The art and science of health promotion*. Hilton Head, SC.
Wetzler, H.P., & Cruess, D.F. (1985). Self-reported physical health practices and health care utilization: Findings from the national health interview survey. *American Journal of Public Health*, **75**, 1329-1330.
Wiedenfeld, S.A., O'Leary, A., Bandura, A., Brown, S., Levine, S., & Raska, K. (1990). Impact of perceived self-efficacy in coping with stressors on components of the immune system. *Journal of Personality and Social Psychology*, **59**(5), 1082-1094.
Wolinsky, F.D. (1978). Assessing the effects of predisposing, enabling, and illness-morbidity characteristics on health service utilization. *Journal of Health and Social Behavior*, **19**, 384-396.
Yen, L.T., Edington, D.W., & Witting, P. (1991). Associations between health risk appraisal scores and employee medical claims costs in a manufacturing company. *American Journal of Health Promotion*, **6**, 46-54.

Chapter 5

Health Care Cost

R. William Whitmer

Never has there been so much attention directed to the cost of medical care and the development of ways to moderate or reduce annual increases. In 1992, about 14 cents of every dollar spent in the United States went for medical care (News Briefs, 1993a). This represented an estimated average cost of $2,953 per person (Data Watch, 1991). When examining employee medical benefits, the average cost per employee was $3,968, which amounted to an average of 42% of net company profits (Data Watch, 1992). For 1992, the increase in indemnity plans was 14.2% and 8.8% for HMOs, which was a combined increase of 10.2% (News Briefs, 1993b). During the past 10 years, the annual increase in medical expenses was over twice the increase in general inflation (Issues and Trends, 1992). The total cost for medical care in 1992 was about $817 billion and was predicted to increase to $940 billion for 1993. Increases over the next 5 years are projected at 12% to 13% per year (News Briefs, 1993b). If past trends continue, by the year 2030 medical care will account for about 28% of the Gross Domestic Product (Waldo, Sonnefeld, & Lemieux, 1991). The economy cannot tolerate this level of expenditure. The present system cannot continue to operate as it has in the past.

Keeping the cost of medical care in the U.S. at a reasonable level and providing medical care for all Americans has been a social and ethical problem for a long time. It has now become a political issue called health care reform. Terms such as "managed competition" and "global budgeting" have appeared in the lexicon. Global budgeting is defined as the establishment of maximum charges that are paid for specific medical services, treatments, and procedures. The concept of managed competition as originally introduced establishes large regional, nonprofit organizations that collect premiums from many large corporations, governmental employers, small businesses, and the self-employed. They negotiate on behalf of the group with health care providers, such as "super" HMOs or large physician/hospital networks, to purchase medical care based on exclusivity and high-volume purchases. All employees are provided with a basic medical plan that may or may not provide for physician and hospital choices and may include some level of rationing. Individuals may select more comprehensive medical plans, some of which preserve physician and hospital choices and provide unlimited care. In this case, the individual pays the cost of the difference between the two plans with nondeductible after-tax dollars. No one is excluded due to preexisting medical conditions, and the poor and unemployed are provided with the same plan.

An advantage of health care reform is the provision of universal coverage with no one denied coverage due to pre-existing conditions. There should also be less expensive administrative work. The disadvantages are the loss of physician choice, some level of rationing, and potential waiting lists for elective surgery and high-tech diagnostic testing. Some authorities estimate that health care reform will require up to $150 billion annually in new taxes, some of which may be raised through increased "sin" taxes on alcohol, tobacco, and firearms.

This summary of the current and future status of the U.S. health care system and its attendant costs reflects some uncertainty and perhaps insecurity. Conversely, with interest in the reduction of utilization and expenses at an all-time high, opportunities are presented to accurately examine health promotion as an effective and creditable methodology for reducing utilization and medical expenses. Based on certain perceptions about health promotion cost savings, considerable work needs to be done. This is reflected in a recent survey where 41% of responding executives indicated they were not convinced that health promotion programs provide cost savings. The same study indicated that 40% of the respondents were concerned about the expenses required to fund health promotion (Data Watch, 1992).

The Definition of Health Promotion

Nearly all the proposals for health care reform mention prevention. Several place high priority on the concept through clinical preventive measures such as inoculations for childhood diseases, AIDS awareness, reduction of lead poisoning in children, and selected prenatal care. However, none provide specific protocols or details for worksite health promotion programs.

The exact percentage of employers providing health promotion programs is uncertain. Reports range from 24% to 81% (President's Council on Physical Fitness and Sports, 1989; Office of Disease Prevention and

Health Promotion, 1993). This wide variation may be due to nonstandardized definitions of health promotion. A few health promotion programs are comprehensive, including regular medical screening, physician referral, on-site intervention programs, fitness centers, etc. Many others are simple and impersonal, involving only a newsletter, posters at the worksite, or an educational paycheck stuffer. Should both of these examples be called health promotion programs? Possibly. Are they comparable in their potential to reduce medical expenses? Definitely not.

This wide variation in the reported number of worksite health promotion programs illustrates one of the first problems encountered when trying to determine cost-effectiveness and other benefits of health promotion. There is no universally accepted definition of a worksite health promotion program. Health promotion professionals have thus far not defined a basic protocol, which if implemented properly would have the greatest potential to be cost-effective. There are several questions that may serve as a framework in the development of the most cost-effective protocol(s):

- If a medical screen is used to identify risk factors, how can participation be increased to the 90% range or higher and maintained over an extended period?
- Should the frequency for rescreening be different depending on the number and severity of risk factors? If so, what is the most cost-effective schedule?
- What methodology is used to ensure that those with major risk factors are referred to and come under the care of a primary care physician?
- What are the standards that determine those intervention programs with the greatest potential for long-term lifestyle change, and which are the most cost-effective?

This is not to suggest that successful worksite health promotion programs will not have individual and unique characteristics. They will and should, but health promotion in general will have the greatest impact on medical expenses if basic and successful methodologies, techniques, and protocols can be developed, replicated, and perfected.

Factors Influencing Medical Expenses

It is essential to recognize that employee medical expenses can be influenced by a number of factors and is therefore a complex issue. Factors that can influence medical costs are

- health promotion,
- changing from indemnity to managed care plans,

- other plan-redesign activities that include cost shifting and other incentives for reduced utilization,
- joining coalitions that emphasize direct contracting based on high-volume purchasing, and
- managed programs for workers' compensation.

When attempting to evaluate the cost-effectiveness or cost-benefit ratio of any worksite health promotion program, it is necessary to identify nonhealth promotion factors that may suggest favorable cost impact from health promotion efforts.

An example may help illustrate this point. An employer starts a comprehensive health promotion program that includes medical screening, physician referral, and intervention programs. Studies suggest that the medical screen may identify problems that initially could increase physician visits and prescriptions for medications (Edington, 1992). Concurrent with the start of the health promotion program, the employer switches from an indemnity, fee-for-service medical plan to managed care, where fees are based on capitated rates. The employer also starts a managed pharmaceutical plan that provides employee incentives for mail-order, generic medications. The end result is that medical expenses are reduced. Is the cost-effectiveness due to the health promotion component, the plan redesign, the new drug plan, or all three? Should health promotion receive the credit because hypertension is diagnosed and brought under control, thus perhaps preventing a long hospital stay due to stroke? How is credit assigned to the reduction in the cost of primary health care through capitated charges? Will the increase in the number of prescriptions caused by diseases discovered through the medical screen be offset by the discounts of the mail-order, generic-oriented pharmacy plan?

Reviewers of the literature on the impact of health promotion to reduce medical expenses usually do not recognize or analyze the role of these nonhealth promotion efforts. I do not mean to deemphasize health promotion as a bona fide and realistic approach to cost reduction; rather, I want to suggest that other cost-containment activities should be included in the analysis.

Health Care Cost Reduction

Increasing reports from the literature suggest health promotion is *cost-effective* and usually provides a positive *cost-benefit ratio*. Both of these techniques will be considered.

Cost-Effectiveness

The literature indicates that the cost-effectiveness of worksite health promotion has been measured in several ways, including the following:

• **The impact on overall medical expenses.** This is the measurement of greatest interest to employers who make the decision to pay for health promotion programs. It is also one of the most difficult to calculate. In order to have maximum validity during the entire study period, all other factors that could influence outcome such as managed care, indemnity plan, cost shifting, and other plan-design factors must remain constant. If cost per employee is compared with other employee groups, the groups must be randomized and statistically similar.

• **The correlation of the number and type of risk factors with medical expenses.** This is one of the more definitive methods to determine cost-effectiveness. The assumption is that the greater the number of risk factors, the more often the person is ill, the more severe the diseases, and the greater the expenses. Because of the potential for risk-rating or variable premiums in paying for health insurance, this method should be thoroughly investigated. Of special interest is the impact of the "uncontrollable" risk factors such as heredity, gender, race, and age.

• **The reduction of measurable risk factors.** This measurement involves repeated screening evaluations so improvements can be measured from one testing to another. Determination of cost-effectiveness is somewhat speculative, however. For example, if severe hypertension is discovered and this is brought to normal levels by medication and/or behavioral change, has a stroke been prevented? Has premature death been delayed? How are these proven? What value is assigned? This frequently used method is based on the assumption that reduction of risk factors results in fewer clinical diseases and medical problems and ultimately in less expenditure.

Pelletier (1991) analyzed and reported on 24 health promotion program studies published from 1980 through 1991. The studies evaluated health and, in some cases, cost benefits of health promotion programs. All 24 demonstrated positive effects for health benefits, and all studies analyzed for cost benefits indicated positive return. A second report by Pelletier (1993) indicated that from 1991 to 1993, 27 new studies were conducted and published. Of the 27, all reported positive health outcomes. Of the studies analyzed for cost-effectiveness, all but one indicated a positive return.

Yen, Edington, and Witting (1991) reported on health promotion efforts for Steelcase, Inc., a manufacturing firm with 7,433 employees. In 1985, health risk appraisals (HRA) were completed by 4,276 (57%) of the employees. Several medical benefits plans were available, including managed care (HMO) and an indemnity plan. Since managed care plans are purchased on a capitated or cost-per-person, per-month basis, the individual cost per medical case was not available. For this reason the study group was reduced to those 1,990 individuals who were in the indemnity plan. Of these, 199 were eliminated because of the possibility of having medical coverage through other sources. This resulted in a study group of 1,838 members (24%) of the work force who completed an HRA and had indemnity medical coverage from 1985 to 1988.

Eighteen risk factors were evaluated. Twelve were self-reported, three were clinically measured, and three were the result of computer projections. The risk factors were divided into three categories: lifestyle habits, psychological perceptions, and health risks.

Medical claims were added together to determine the total cost per employee per year. Expenses for pregnancy were excluded. Based on the annual cost for the 3-year period, an average annual cost per employee was determined.

Of the 1,838 employees, 306, or about 16%, had no medical claims. At the other extreme, 10% of the employees utilized 64% of the medical care dollars. The remaining 74% of the employees incurred only 36% of the medical expenses.

As shown in Figure 5.1, those with no risk factors spent an average of $190 per year. Those with one risk factor used $360 annually. Persons with two to three risk factors had annual medical expenses of $542, and those with four to five had expenses of $718. Employees with six or more risk factors had annual expenses of $1,550. Therefore, it is concluded that "as the number of high risk categories per employee increased, so did the likelihood of having high-cost status" (Yen, Edington, & Witting, 1991, p. 51).

Henritze, Brammell, and McGloin (1992) reported on a study for Coors that produced similar findings. This report also indicated that as risk factors increased, medical benefit costs went up accordingly. The major difference was that Coors employees seemed to incur considerably higher expenses per risk factor.

Coors had 1,320 employees at a single location. The medical screening included blood pressure, finger-stick total cholesterol, height, weight, body-mass index, smoking, and physical activity history. If the cholesterol was in excess of 200 mg/dl, a lipid profile and glucose test were done. If systolic blood pressure was over 140 mm/Hg and/or diastolic blood pressure over 90 mm/Hg, then the blood pressure test was repeated.

The medical screen was voluntary and 692 (52%) participated. A second medical screen was offered and 499 (37%) of the total employee population participated. Approximately 77% of participants were male with the average age about 43 years old. It was found that 91% had one or more risk factors and 33% had three to five.

An attempt was made to correlate medical claims with risk factors for 1989. Statistics are provided for men and women. Figure 5.2 shows that for men there was no statistical difference in medical claims among those with one to three risk factors. One risk factor reflected annual costs of $900, and two to three risk factors cost $850. Those with four risk factors had $1,100 in costs, and five risk factors cost $3,100. Figure 5.2 also shows

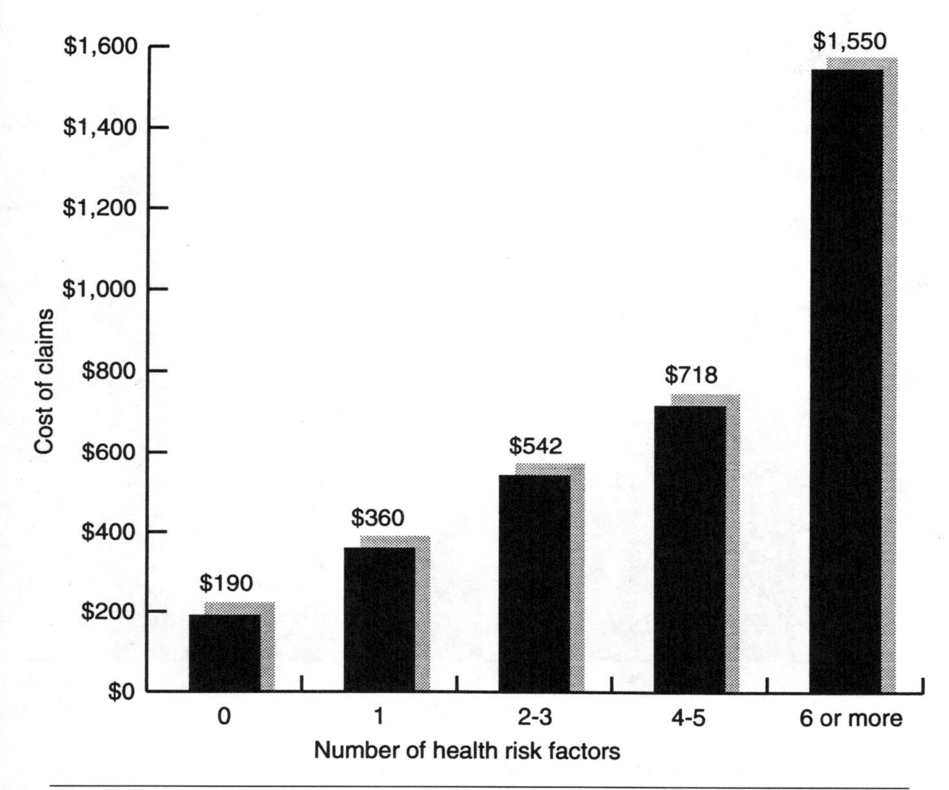

Figure 5.1 Average employee medical claims per number of health risk factors at Steelcase, Inc.
Note. Data from "Associations Between Health Risk Appraisal Scores and Employee Medical Claims Cost in a Manufacturing Company" by L. Yen, D. Edington, and P. Witting, 1991, *American Journal of Health Promotion*, **6**(1), pp. 46-54.

that women with one risk factor had medical claims of $1,050. Two risk factors were associated with $1,200 in claims. Three utilized $1,900, and four accounted for $1,800 in medical claims. No women had more than four risk factors.

Naas (1992) analyzed and reported on a more specialized second study from Coors that included the impact of a cardiac rehabilitation program. Employees recovering from heart attack or heart surgery were required to participate in a 12-week program supervised by a physician. The program included aerobic exercise classes, nutrition, and stress-management classes. A striking difference was noted in time off following a cardiovascular illness before and after the program was implemented. Average time off before the program was 7.3 months. After the program was introduced, time off was reduced to 2.1 months. It is estimated the program has saved $2.3 million in lost wages and $1.9 million in rehabilitation expenses and cost avoidance.

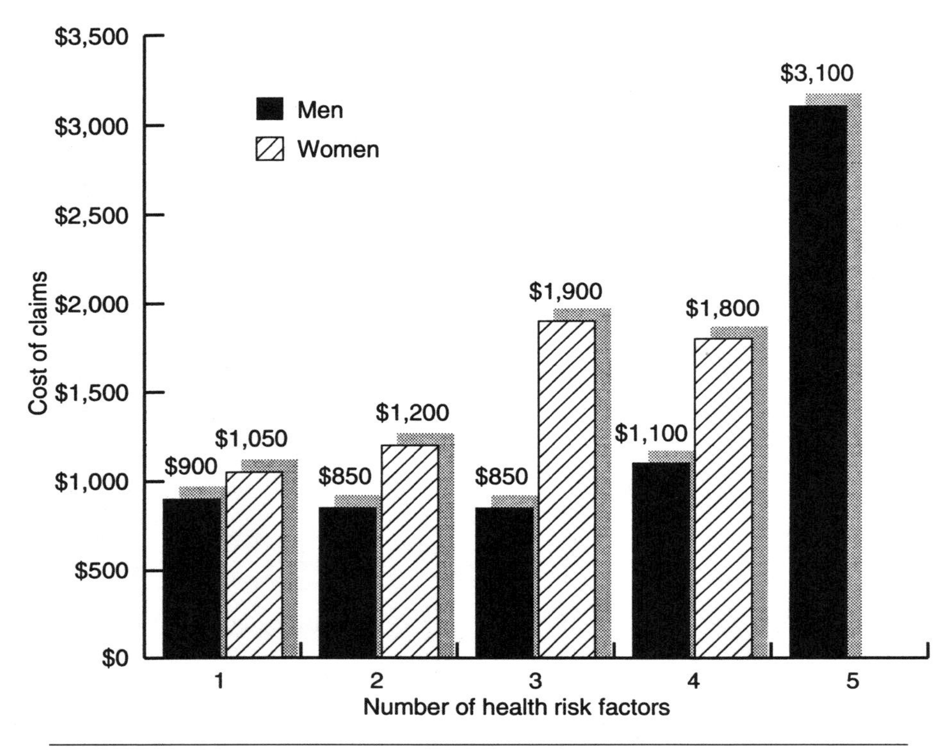

Figure 5.2 Average employee medical claims per number of health risk factors (men vs. women) at Coors Brewing Company.
Note. Data from "Lifecheck: A Successful, Low Touch, Low Tech, In-Plant Cardiovascular Disease Risk Identification and Modification Program" by J. Henritze, H. Brammell, and J. McGloin, 1992, *American Journal of Health Promotion*, **6**(4), pp. 129-136.

Based on real and anticipated reduction in medical expenses, reduced sick leave, and increased productivity, the overall wellness programs save about $3 million annually. Annual expenses are $1 million. Health promotion efforts have enabled this employer to achieve a cost/benefit ratio of $5.50 saved for each dollar invested.

Health promotion activities of the 1,200-employee Canada Life Assurance Company are reported by Shephard (1992). The company has a fitness center that is used by 13% of the employees. To estimate the impact of the fitness program, a study was conducted that compared per capita medical claims with another statistically comparable 800-employee insurance company that did not provide a fitness program. Over a 1-year period, per capita medical expenses for the control group increased 25%, from $170 to $229 (1990 Canadian dollars). The study group with the fitness center maintained per capita costs at $170 (1990 Canadian dollars) for zero increase.

The initial projected cost-benefit ratio was $6.85 for each dollar invested. This was based on 20% employee participation. Since only 13% participated, the cost-benefit ratio is estimated at about half this amount ($3.43).

Harvey, Whitmer, Hilyer, and Brown (1993) reported on a 5-year, National Institutes of Health–funded health promotion program for the employees of the city of Birmingham, Alabama. This study is unique in that over the 5-year period, an average of 95% of the 4,000 employees participated annually in a comprehensive medical screen. This was accomplished by making the medical screen a prerequisite for enrollment in the medical benefits plan. Following the medical screenings, 13% were found to have significant risk factors and were provided individual health care consultations. It was determined that about 40% of this group did not have a physician/patient relationship. When permitted, assistance was offered in providing referral to a primary care physician. Intervention programs were offered over the 5-year period.

Cost-effectiveness was determined by comparing annual per-employee cost for medical benefits with the per-employee average for the state of Alabama. Table 5.1 shows that in 1985, the per-employee cost was $2,047, or $397 above the state average. In 1990, the per-employee cost was $2,075, or $922 below the state per-employee average. From 1985 to 1990 the city had no increase in the cost of medical benefits, while average per-employee medical expenses doubled for other employers across the nation.

In another report on the same study Whitmer (1992) calculated a cost-benefit ratio by comparing actual medical expenses with a 5-year forecast for the period of 1985 to 1990. The forecast, compiled by a national medical

Table 5.1
City of Birmingham Per-Employee Medical Expenses Compared With State of Alabama Per-Employee Medical Expenses

Period	Alabama average cost/employee	Birmingham cost/employee	Per-employee difference between city and state
1985	$1,650	$2,047	$397
1986	1,770	2,084	314
1987	1,997	1,943	(54)
1988	2,315	2,027	(288)
1989	2,595	2,053	(542)
1990	2,997	2,075	(922)

Note. Adapted from "The Impact of a Comprehensive Medical Benefit Cost Management Program for the City of Birmingham, Alabama: Results at Five Years" by M. Harvey, W. Whitmer, J. Hilyer, and K. Brown, 1993, *American Journal of Health Promotion*, **7**(4), pp. 296-303.

benefits consulting firm, was based on the trend in increases from 1974 to 1984. Comparison of actual expenses with this forecast indicated the city saved $31 million over the 5-year period. After the $3 million spent on the health promotion program was deducted from the savings there was a cost-benefit ratio of $9.33.

Leutzinger, Goetzel, Richling, and Wade (1993) reported on health promotion activities at Union Pacific Railroad. This employer has 28,000 employees of which 20,000 (71%) have access to health promotion programs at 59 sites. In 1990, medical expenses were more than $360 million. About 1% of this amount was spent on health promotion activities.

A study was conducted on all employees under 65 years of age who had medical coverage through the indemnity plan. The purpose was to determine what percent of the medical expenses were caused by unhealthy lifestyles. It was found that 28% ($87 million) of the medical expenses were lifestyle-related. This compared with an average of 15% for other employers. The $87 million in excess medical expenses added an extra $439 in costs for each plan member in 1990. Employees and spouses accounted for 73% of the excess cost and dependents under 17 years of age incurred the remaining 27%. Stress was the most prominent negative influence, causing more than $35 million in excess medical expenses.

Bertera (1991) reported on a study by the E.I. Du Pont de Nemours Company that was an evaluation of health risk appraisal and physical examination data over a 4-year period. Nearly 46,000 employees were involved. The influence of seven risk factors on absenteeism and medical expenses were determined. Those with zero to two risk factors were considered low risk, and those with three or more, high risk. High-risk employees were more often male, married, nonwhite, and paid hourly. When comparing low-risk with high-risk employees, the high-risk employees had excess medical expenses caused by the following factors: smoking ($960), obesity ($401), excess alcohol consumption ($389), elevated cholesterol ($370), high blood pressure ($343), lack of seat belt usage ($272). Lack of exercise was not a significant factor.

When absenteeism trends at 71 Du Pont worksites with health promotion programs were compared with 20 control worksites without programs, it was found that absenteeism at the health promotion sites declined by 14% compared to a 5.8% decline at worksites without programs. Based on annual health promotion expenses of $40 per employee, a cost-benefit ratio of $2 for every $1 spent was reported.

Naas (1992) analyzed and reported on the experience of Baker Hughes, Inc., which has 17,000 employees and is involved in the manufacture of oil field equipment. The company provides a basic voluntary medical screen with 65% participation. Employees were risk-rated based on cholesterol, triglyceride, blood pressure, and weight. If they were normal in three of the four areas, they were provided $100 annually to pay out-of-pocket medical expenses. Smokers paid an extra premium of $10 monthly. This resulted in a plan that ranged from a $120 penalty to a $100 reward for a positive lifestyle. Total medical expenses for 1991 were expected to be $50 million. Actual expenses were $43 million.

Bernacki (1987) and Bernacki and Baun (1984) reported on the effects of the Tenneco Company's health promotion program, which provides a major fitness center for the 3,600 employees who work at corporate headquarters. The results indicate that female employees who exercise have average medical claims $900 less than female nonexercisers. Male exercisers have $500 less in medical expenses than male nonexercisers.

Nonexercising females were absent from work 69 hours a year compared to 47 hours for the exercisers, which is a 31% reduction. The difference for males was not as impressive. Male exercisers were absent 25 hours compared to 30 hours for nonexercisers—a 16% reduction for the exercisers.

Connors (1992) analyzed and reported on the health promotion program for the 18,000 employees of GE Aircraft. Approximately 5,500 employees (30%) are members of the company-owned fitness center. A 3-year study indicated that medical claims for fitness center members declined from $1,044 annually to $757, a 27% decrease. Medical claims for nonmembers went from $773 to $941 per year, a 17% increase. The greatest change in expenses seemed to be from shorter hospital stays for members. A specific cost-benefit ratio was not provided but company officials "feel" the savings were about equal to fitness center expenses and program costs.

Cost-Benefit Ratio

Attempts to compare calculated savings from health promotion programs with actual program expenses provide a cost-benefit ratio. In some cases actual costs are compared to "anticipated" or forecast costs. This methodology raises questions: Who prepared the forecast? What is the basis of the forecast? Are there unanticipated factors that can influence the forecast? Another methodology measures "cost avoidance," which compares expenses prior to a health promotion program with post-program expenses. While there can be extenuating influences, this is a reasonable method to determine a cost-benefit ratio.

Table 5.2 presents a summary of reported cost-benefit ratio data. The Canada Life Assurance cost-benefit ratio was $3.40 and based on calculated or projected savings. It includes projected impact on turnover, productivity, and medical claims (Shephard, 1992). The city of Birmingham cost-benefit ratio was $9.33, which is based on comparison with a forecast calculated as an extension of the previous 10 year trend of increases (Whitmer, 1992). The Coors cost-benefit ratio of $5.50 was calculated from real and anticipated reduction in medical costs, reduced sick leave, and

Table 5.2
Cost-Benefit Ratio

Company	Dollars saved for each dollar invested
Canada Life Assurance Corp. (1990 Canadian dollars)	$3.40
City of Birmingham	9.33
Coors	5.50
Du Pont	2.00
GE Aircraft	-0-
General Foods	3.50
Indiana Blue Cross/Blue Shield	2.51
Univerity of Michigan Worker Health Program	2.34

increased productivity (Henritze et al., 1992). The Du Pont program saved $2.00 for every dollar spent and was calculated based on reduced absenteeism (Bertera, 1991). GE Aircraft reports that program expenses equaled savings (Naas, 1992). The General Foods cost-benefit ratio was $3.50 and based on comparing participants with nonparticipants in the areas of absenteeism, fitness, blood pressure, cholesterol, proper weight, and use of tobacco (Wood, 1989). The Blue Cross/Blue Shield program had a cost-benefit ratio of $2.51, based on compared risk factors, absenteeism, and medical costs among participants and those who did not participate in health promotion programs (Mulvaney et al., 1987). The University of Michigan saved $2.34 for every dollar spent when comparing medical expenses for hypertensive individuals who did and who did not participate in a health promotion program (Foote & Erfurt, 1991).

Medical Screen Participation

Increasing evidence suggests that as the number of risk factors increase, so do medical expenses. For this reason, considerable interest exists in identifying those who are at high risk. The most common way to accomplish this is through medical screening. Nearly all the medical screening data reported in the literature is based on voluntary participation (Henritze et al., 1992; Jones et al., 1991; Naas, 1992; Shephard, 1992; Yen et al., 1991). This may not be an effective method to identify those who are high risk because it is thought that when a medical screen is voluntary, those who participate have fewer risk factors and may be more healthy (Edington, 1992). Those with multiple risk factors are often intimidated by the screening and therefore decline. This presents a case of adverse selection and prevents the identification of those at high risk. There is good evidence that in most cases a small number of employees use a very large percentage of the medical benefit dollars. One report indicates that only 1% of all employees utilize 30% of medical benefits (Issues and Trends, 1993). Another report suggests that 10% of employees use 64% of the medical benefit dollars (Edington, 1992). It is suspected that those who decline voluntary medical screening may make up a large percentage of the small group that creates the majority of medical claims.

Several methods may be considered to maximize employee participation in the medical screen:

- Make the medical screen a prerequisite for enrollment in the medical benefits plan.
- Make the employee financial contribution for medical benefits considerably lower if there is participation in the medical screen.
- When the medical screen is conducted, provide every employee with an appointment time. While there is no requirement and the employee may decline, the inference is that the employer expects the employee to participate.

When the medical screen was made a prerequisite for medical benefits, 95% participation in a medical screen over a 5-year period was achieved (Harvey et al., 1993). Unpublished experiences of the author with a privately owned Coca-Cola Bottling Company and a privately owned Pepsi-Cola Bottling Company resulted in 90% participation when the third technique described above was used.

Participation in a health promotion program can be divided into three groups: those who participate in a medical screen, those who seek medical care for problems identified through the medical screen, and those who participate in lifestyle change programs. This concept elevates the role of the primary care physician and recognizes the necessity for therapeutic management of selected risk factors.

Spouses and Dependents

Edington (1992), observed that spouses utilize about 35% of the health care dollars and other dependents about 25%. This confirms the notion that covered spouses/dependents create more medical expenses than employees. Despite this observation, there are few reports describing health promotion experiences with covered spouses/dependents. No reports could be found indicating specific cost savings caused by health promotion with this group.

As health promotion programs mature, involvement of spouses covered by the medical benefits plan will become popular. Presently, the delivery of worksite health promotion programs to children and adolescent dependents is infrequent due to a variety of factors.

The Role of the Physician in Cost Containment

A logical place for health promotion to be practiced is in the physician's office. Patients often follow instructions given by their personal physician. This presents an excellent opportunity for serious in-depth health promotion counseling to take place. Unfortunately, such counseling is rare. Physicians are usually not trained in the area of health promotion or in the psychology of behavioral modification. Usually, their training and prime interests are diagnosing and curing medical problems and diseases after they occur. Additionally, most physicians have busy schedules and health promotion counseling is time consuming.

Unfortunately, medical benefit plan design usually does not allow for payment of health promotion services. Despite this situation, some physicians do counsel patients to quit smoking, lose weight, or cut back on high-cholesterol foods, but this is usually secondary to treatment and cure.

The Role of the Insurance Company in Cost Containment

The question has been raised about insurance companies paying part of all health promotion program expenses. This assumes there will be fewer illnesses and therefore reduced expenses. This seems to be a situation that could benefit the insurance company, but it usually does not. Most employers who provide health promotion programs are self-insured. This means all medical expenses are paid by the employer from net profits. There may be an "insurance company" involved that serves as the third party administrator (TPA). It is the function of the TPA to do preadmission certification, utilization review, and coordination of benefits, and to provide reinsurance, receive bills, write checks, and provide the employer with reports. For these services the TPA is paid a fee. All medical claims are paid from reserve funds provided by the employer from net profits.

Although it may seem logical that health insurance companies should fund health promotion, it is not a realistic request. With self-insured companies, it is the employer who will benefit financially from fewer medical problems and lower medical benefits expenses.

Companies too small to be self-insured usually buy medical coverage through a "trust" or "pool." These small employers pool the premiums paid and spread the expenses over the entire group. Under these conditions, it may be difficult for an employer to justify the expenses of a health promotion program because they are "community rated." With community rating, individual medical expenses are the same for all employees, depending on the illnesses and utilization of everyone who is in the insured group. This could eliminate the incentives to provide worksite health promotion based on the assumption or anticipation of reduced utilization and lower medical expenses.

Conclusion

For the number of recognized, mature health promotion programs in the U.S., few have been analyzed and reported in the scientific literature. One explanation may be that executive management has a notion that employees who take care of their health are better employees and this is sufficient justification for the program. Another factor may be that research can be expensive and up to now health promotion research does not have the same priority as clinical medicine research.

Perhaps the principal reason for the lack of reported studies is that the accurate measurement of the impact of worksite health promotion is not an exact science. The main problem facing researchers is that the procurement, administration, and delivery of medical services is a highly complex, multidimensional activity. Human resource executives are constantly changing this procurement, administration, and delivery system to reduce utilization and costs. This creates an environment in which it is very difficult to determine the precise impact of health promotion.

Despite these problems, progress is being made. Health promotion as a discipline is only several decades old. During this brief time methodologies have been developed that result in up to 95% participation in medical screens. Rick factors are being positively correlated with medical claims. Increasing numbers of employers are beginning to accept and provide health promotion as part of the corporate culture.

Following are several areas that require research effort:

1. **Corporate Impressions.** Most health promotion programs are purchased by corporate executives. Research is needed to determine how to enhance the perception of health promotion and make it a part of the medical benefits plan.
2. **Cost-Effectiveness.** Methodologies are needed to more accurately determine the specific economic impact of health promotion within the changing environment of the workplace. Prospective studies are required to determine short- and long-term medical expenses based on specific risk-factor combinations and determine if medical expenses are reduced as risk factors are modified or eliminated.

3. **Protocols.** Methodologies need to be developed to increase participation in medical screening. Emphasis must be placed on getting the primary care physician into the health promotion network by referring all high-risk participants for proper diagnosis and treatment. Effective incentives for participation in intervention programs should be identified and evaluated.
4. **National Research Clearinghouse.** Even though health promotion research dollars are not plentiful, research needs to be coordinated by a central organization that could assess research needs and assign priorities so as to minimize duplication and enhance outcomes.

References

Bernacki, E., & Baun, W. (1984). The relationship of job performance to exercise adherence in a corporate fitness program. *Journal of Occupational Medicine*, **26**(7), 529-531.

Bernacki, E. (1987). Can corporate fitness programs be justified? *Fitness in Business*, **1**(5), 173-174.

Bertera, R. (1991). The effect of behavioral risks on absenteeism and health care costs in the workplace. *Journal of Occupational Medicine*, **33**(11), 1119-1124.

Connors, N. (1992, March). Wellness promotes healthier employees. *Business and Health*, pp. 66-71.

Data Watch (1991, December). A glimpse into the future (U.S. per capita health spending). *Business and Health*, p. 14.

Data Watch (1992a, July). A snapshot of executive poll results. *Business and Health*, p. 14.

Data Watch (1992b, September). Reasons for not offering health promotion activities. *Business and Health*, p. 14.

Edington, P. (1992). Is it possible to simultaneously reduce risk factors and excess health care costs? *American Journal of Health Promotion*, **6**(6), 403-406.

Foote, A., & Erfurt, J.C., Jr. (1991). The benefit to cost ratio of work-site blood pressure control programs. *Journal of the American Medical Association*, **256**(10), 1283-1286.

Harvey, M., Whitmer, W., Hilyer, J., & Brown, K. (1993). The impact of a comprehensive medical benefit cost management program for the city of Birmingham, Alabama: Results at five years. *American Journal of Health Promotion*, **7**(4), 296-303.

Henritze, J., Brammell, H., & McGloin, J. (1992). Lifecheck: A successful, low touch, low tech, in-plant cardiovascular disease risk identification and modification program. *American Journal of Health Promotion*, **6**(4), 129-136.

Issues and Trends (1992, March). Cost shifting burden totals $17.2 billion. *Business and Health*, p. 22.

Issues and Trends (1993, March). Small group spends majority of health care dollars. *Business and Health*, p. 18.

Jones, R., Sankoff, J., Wolff, C., & Bowers, T. (1991). Employee health promotion at a university medical center: A pilot project. *American Journal of Health Promotion*, **6**(1), 7-9.

Leutzinger, J., Goetzel, R., Richling, D., & Wade, S. (1993, March). Projecting the impact of health promotion on medical costs. *Business and Health*, pp. 38-44.

Mulvaney, D., Gibbs, J., Reed, W., Grove, D., & Skinner, T. (1987). Staying alive and well at Blue Cross and Blue Shield of Indiana. In J. Cepatz (Ed.), *Health promotion evaluation: Measuring the organizational impact.* Washington, DC: National Wellness Institute.

Naas, R. (1992, November). Health promotion programs yield long-term savings. *Business and Health*, pp. 41-47.

News Briefs (1992, February). Health care spending to consume 14% of GNP in 1992. *Business and Health*, p. 10.

News Briefs (1993a, February). Health care costs rose sharply in 1992. *Business and Health*, p. 10.

News Briefs (1993b, March). Employers benefits spending moderated in 1992. *Business and Health*, p. 10.

Office of Disease Prevention and Health Promotion. (1993). *1992 National Survey of Worksite Health Promotion Activities*. Washington, DC: Author.

Pelletier, K. (1991). A review and analysis of the health and cost effectiveness outcome studies of comprehensive health promotion and disease prevention programs. *American Journal of Health Promotion*, **5**(4), 311-315.

Pelletier, K. (1993). A review and analysis of the health and cost effectiveness outcome studies of comprehensive health promotion and disease prevention programs at the worksite: 1991-93 update. *American Journal of Health Promotion*, **8**(1), 50-61.

President's Council on Physical Fitness and Sports (1989). *Survey to determine the percentage of employers providing health promotion programs to employees.*

Shephard, R. (1992). Twelve years experience of a fitness program for the salaried employees of a Toronto life assurance company. *American Journal of Health Promotion*, **6**(4), 292-301.

Waldo, D., Sonnefeld, S., & Lemieux, S. (1991). Health spending through 2030: Three sections. *Health Affairs (Millwood)*, **10**(4), 231-242.

Whitmer, W. (1992, March). The city of Birmingham's wellness partnership contains medical costs. *Business and Health*, pp. 60-66.

Wood, A. (1989). An evaluation of lifestyle risk factors and absenteeism after two years of a worksite health promotion program. *American Journal of Health Promotion*, **4**(2), 128-133.

Yen, L., Edington, D., & Witting, P. (1991). Associations between health risk appraisal scores and employee medical claims cost in a manufacturing company. *American Journal of Health Promotion*, **6**(1), 46-54.

Chapter 6

Health Behaviors and Risks Appraisal

D.W. Edington

The assessment of health behaviors and risks is commonly used in health promotion to place individuals into risk categories and for program and case management strategies. Other major contributions of behaviors and risks assessment include program promotion, awareness, education, and evaluation.

A recent application of health behaviors and risks assessment technology has been to associate behaviors and risks with the costs of health care. In most cases thus far reported, high-risk behaviors have been associated with high costs for health care. These high-risk/high-cost and low-risk/low-cost relationships can be misleading because not all high-risk individuals translate into high cost or low-risk into low cost. The analysis is complicated by the skewed nature of the cost distribution: 10% of the population accounts for 60% (multiple years) to 80% (single year) of the costs. Clarification of the association between risks and costs will require a new generation of data analyses.

The purpose of this chapter is to briefly review the association of health behaviors and risks with health status and economic indicators, the use of incentives in promoting behavior and risk changes, and the economic impact of those changes. Only recently have the data relating health behaviors and risks to economic impact become available. Therefore, most of the reports reviewed in this chapter were published in 1990 or later.

Association of Health Behaviors and Risks With Health Status

Lewis Robbins, MD, must receive much of the credit for promoting the concept of lifestyle behaviors and health risks as precursors of future disease status (Robbins & Hall, 1970). The Carter Center Health Risk Appraisal (derived from the original Centers for Disease Control HRA) identifies 28 precursors to 62 diseases. This precursor concept has spawned the professional field we now know as disease prevention, health promotion, or wellness.

The original work by Robbins and Hall also gave rise to the landmark report by the Surgeon General, *Promoting Health/Preventing Disease: Objectives for the Nation* (1979, 1980) and the subsequent *Healthy People 2000: National Health Promotion and Disease Objectives for the Nation* (U.S. Department of Health and Human Services, 1990). This latter document now provides the basis for most public and private health goals for the nation. The health behaviors and risks identified in these documents and the stated goals could be widely accepted as benchmark indicators of success for any worksite program (Muchnick-Baku & McNeil, 1991).

Health behaviors and risks appraisals (HRAs) are generally accepted as the single most convenient method of collecting and tracking health behaviors and risks in any size population. Edington and Yen (1992b) provide a comprehensive review of the reliability, validity, and effectiveness of health risk appraisals. In summary, the review indicates the HRA is incredibly accurate in predicting future (10- or 20-year) mortality (Wiley, 1980; Kramer, Wiley, & Camacho, 1981; Foxman & Edington, 1987; Kannel & McGee, 1987; and Gazmararian, Foxman, Yen, Morgenstern, & Edington, 1991). An earlier, more comprehensive review of HRA technology was completed in 1987 (National Center for Health Services Research). The consensus panel concluded that the HRA technology had considerable merit and that additional and updated scientific documentation was needed. Also, the experts agreed that the HRA was an excellent tool for doing epidemiological research but lacked specificity for application to individuals. Finally, there was little evidence that the HRA (1980 version) was an effective behavior modification intervention. In regard to this latter point, these experts missed the fact that not all individuals are high risk nor are they uniformly ready for change, and thus no tool will insure a uniform, effective behavior modification program in a wide population.

Although the HRA technology was developed to assess risk for mortality, it has also been proven effective in assessing risk for future illness or utilization and costs of the health care system (Yen & Edington, 1988). The HRA has been shown to be effective in assessing risk for breast cancer (Imrey, Williams, Schmale, Imrey, & Moll, 1983), cardiovascular disease (Kannel & McGee, 1987; Lu, Yen, & Edington, 1993), health care costs (Yen, 1990; Yen, Edington, & Witting, 1991), and overall health status (Dewey & Seehafer, 1987).

Association of Health Behaviors and Risks With Economic Indicators

Concern for the health of individuals should be cause enough to justify corporate health promotion programs. Not only is the appraisal of health behaviors and risks a powerful tool for predicting chances of dying, it is also an efficient way to measure and predict future utilization of the health care system. However, there is an ever-increasing interest in connecting the health of individuals to measures of economic benefit: health care costs, absenteeism, workers' compensation, disability, and other measures of productivity and benefits to the organization.

Articles by Bertera (1991), Yen et al. (1991), and Brink (1986) examine what has become the basis for many of the risk-rating incentive programs currently being introduced in wellness programs throughout the country. The articles find that high-risk behaviors, typically addressed in wellness programs, produce higher medical care claims than low-risk behaviors. Bertera (1991) reported higher average annual illness days (absenteeism) for employees with high-risk behaviors and higher health care costs ranging from $272 to a high of $960. Yen et al. (1991) found differences in health care costs ranging from $190 to $350.

Yen et al. (1991) also reported a linear relationship between the number of high-risk categories and increasing health care costs. For example, employees with zero high-risk categories had an annual average cost to the employer of $190 compared to $1,550 for employees with six or more high-risk categories. A second article by Yen et al. (1992) agrees with Bertera in associating high-risk behaviors with higher absenteeism cost, calculated by multiplying the average annual hourly rate times the days absent.

Conrad, Conrad, and Walcott-McQuigg (1991) and Lynch, Teitelbaum, and Main (1991) caution us about drawing conclusions about worksite health promotion programs and the related medical costs. Conrad et al. (1991) advises researchers and evaluators to be constantly alert to such issues as selection bias, attrition, normal maturation, historical intervention, defusing of treatment, and other factors that may distort the true meaning of program interventions. Lynch et al. (1991) cautions program evaluators against using the means when examining medical costs, especially since costs are notoriously skewed: approximately 10% of a population can account for 58% to 80% of health care cost (Yen, 1990; Yen et al., 1991), depending on the number of years analyzed.

One approach to studying the association of health behaviors and risks to economic indicators can be examined by studying each behavior or risk independently. The limitation of this approach is illustrated in the potential interaction of the behaviors and risks (Yen et al., 1990, 1991, 1992). Nevertheless, the literature is heavily populated by studies reporting outcomes of research focusing on a single behavior. A review of this research follows.

Smoking

Smoking cessation often receives a high priority in worksite and community health programs. The social, psychological, and physical health reasons for not smoking have received considerable attention and are well documented (Office on Smoking and Health, 1989). Analyzing hospital utilizations and costs for tobacco users, Penner and Penner (1992b) found that hospital utilization days and costs were higher in tobacco users in almost all measures. Tobacco users had nearly 75% more hospital days and over 65% more admissions per 1,000 employees. Their chemical dependency and psychiatric days per 1,000 employees were four to five times higher. The study is severely flawed by the authors' failure to control for other behavioral characteristics of the population that would influence hospitalization. Nevertheless, they claim that the use of tobacco can be viewed as a marker for potential high utilizers of the health care system.

Li, Windsor, Lowe, and Goldenberg (1992) and Shipp, Croughan-Minihare, Pettit, and Washington (1992) examined the potential benefits of smoking cessation programs primarily in reducing instances of low birth weight babies and health care costs. The sophisticated analyses by Li and colleagues (1992) show the relative risk (RR) of low birth weight babies from mothers who smoke ranges between two and four. That is, the chances of having a low birth weight baby is two to four times as great in mothers who smoke compared to mothers who do not smoke. Given that approximately 25% of pregnant women smoke, these data would predict a high number of low birth weight babies, who require extensive high-cost medical care, including both direct and indirect costs, over a longer period of time. Smoking cessation programs that are effective during the first trimester would greatly reduce the number of low birth weight babies. Thus, the first trimester is an especially important opportunity for worksite health promotion programs.

Although Li and colleagues (1992) calculated their benefits on a national basis, the same benefits would apply to a local worksite. The benefits to the local population would depend greatly on the percent of smokers in the population, the success of the smoking cessation program, and the cost of the program. Li et al. estimated that the quit rates for smoking cessation programs are approximately 25% and the cost of programs is approximately $30 per person. Shipp and colleagues (1992) determined that the break-even point for smoking cessation programs in prenatal programs is $32 per pregnant woman. The break-even point is influenced by the percentage of smokers in any one group and the estimated success rate.

Analyzing 11 months of paid medical claims data, Penner and Penner (1992) found that smokers had a 12% hospital admissions rate versus a 7% rate for nonsmokers. The smokers' average length of hospital stay was longer and the average outpatient payments were higher. Although the data are in the expected direction, the study had several weaknesses. First, no tests of statistical significance were applied to the study; second, the use of only a few months of medical claims data creates highly suspect data due to variability. Furthermore, the researchers did not separate men and women smokers, age groups, or income levels. Although these weaknesses have been found in similar studies of smokers, the findings continue to be in the same direction.

Nutrition and Weight Control

Siegelman (1991) reviewed the nutrition and weight reduction programs offered by 15 companies. Nearly all of the programs offered some financial incentive, such as significantly reduced programs fees, complete cost reimbursements, or even reduced employee contributions to insurance plans. Siegelman cites an estimate of the Wellness Council of America that approximately 26% of the work force is overweight. Some companies, such as Progressive Insurance Company, reimburse an employee based on the amount or percent of weight loss maintained over a full year. Employees can gain up to $250 reimbursement of their cost. Participants in this program and other health promotion activities earn points toward prizes that can be claimed at the end of the year. Apple Computer offers a similar reimbursement for the $60 fee of a weight reduction course. GE Aircraft offers five different programs to its 18,000 employees, geared to the degree they are overweight and targeted to their different needs. The fees to the employers for the program run approximately $96 per year or 40% of the actual expenditure. GE Aircraft claims the participants in the weight control programs cost $184 less in medical claims per year than nonparticipants.

Hypertension

Suggs, Cable, and Rothenberger (1990) screened nearly 900 employees at 10 worksites for hypertension and oral, breast, rectal, prostate, colon, and testicular cancer. Of the nearly 900 employees, 51 were identified as having hypertension, two with early cancers and four with malignant precursors. Although the total cost of the program was nearly $98,000, the authors suggest it was cost-effective because a late-stage colon cancer or breast cancer treatment has been estimated to cost over $150,000. It is difficult to substantiate claims in a true cost-effectiveness or cost-benefit study. An argument could be made that the cost of the program is in real dollars but the projected savings are in dollars that may or may not be spent in future years.

Exercise

Pavett and Whitney (1990) used personal interviews to report absence and quality-of-life factors in exercisers in two different companies. They report that frequent exercisers had fewer illness-related absences than others and that most of the exercisers used the corporate facilities. The frequent exercisers had at least one less absent day and reported better stress reduction, improved self-concept, higher energy, and better performance. The authors calculated that for a company of 1,000 employees making an average salary of $2,000 per month, the exercise program could result in an $11,000 per month savings ($110 per employee saved per month).

Cholesterol

The National Cholesterol Education Program has advocated cholesterol screening for all adults aged 20 years and older. Froom and Froom (1990) estimate that compliance with this suggestion would result in the identification of nearly 120 million hypercholesterolemics, increasing physicians' workload by 15 office visits per 1,000 adult patients. The mean cost for a cholesterol test was estimated at nearly $17. Additional LDL tests were estimated at $41, and a limited office visit was estimated at $28. Therefore, the total cost of the program was estimated at nearly $12 billion. Additional costs would come in follow-up programs and other means to achieve a high adherence rate. Even with this expenditure, the benefits of the program would be unsure. Possible benefits would include future medical costs avoided, person years saved, productive years saved, and the avoidance of pain and suffering. The authors were also concerned that over half of the laboratories that measure cholesterol have not met the accuracy standards of the National Institutes of Health laboratory standardization panel.

Alcohol

Holder and Blose (1991) reported that "alcoholics" had twice the alcohol-related medical costs of nonalcoholics per year over a 13-year period. The total alcohol-related medical costs for alcoholics were approximately $1,200 per year from 1974 through 1987. While the data are in the expected direction, the authors did not report any standard deviations of their data. It is likely the standard deviations would be significantly higher than the means, which would open their statistical analyses to question. One year later these same authors (Holder & Blose, 1992) presented the 14-year results. Alcoholics were identified by a search of the ICD-9 (Ninth Edition of the International Classification of Diseases) categories. The authors claim that the costs for treated alcoholics decrease within the first year posttreatment and appear to trend downward to a level at or below the immediate pretreatment level. Costs for the untreated groups remain quite elevated for a number of years before declining. The authors claim that the treated alcoholics cost approximately 24% less than untreated alcoholics over a 4-year period. In 1985 dollars, the monthly cost for the untreated alcoholics averaged $201 versus $162 for the treated group. The authors did not report the average nonalcoholic's costs during the same period, but we would assume it to be the same as other populations—less than $50 per month. Holder and Blose report a gradually increasing health care cost prior to identification for alcoholism. Likely, a health promotion program could be most effective in this pretreatment stage to heighten the employee's awareness of the problem. The important dependent variable would become the number of alcoholics identified per year. The goal of the health promotion program would be to decrease the number of unidentified alcoholics.

Drugs

Marini (1991) conducted five interviews to assess corporate initiatives for a drug-free workplace. The author concludes that substance abuse affects one out of every five workers aged 18 to 25. The substance abusers were found to be four times more likely to have accidents, more than twice as likely to have prolonged absences, and five times as likely to file workers' compensation claims and to utilize approximately three times the average health benefits.

Wellness Screenings

In many wellness programs, baseline data are collected through a screening mechanism and connected to follow-up procedures with recommendations made for health behavior change programs. The methods of health screenings vary from program to program. Most screening programs include a combination of physiological measures and questionnaire data. Some of these measures are self-reported whereas others may be checked by a health professional.

Foote and Heirich (1991) and Szymanski, Pate, Dowda, Blair, and Howe (1991) attempted to separate the basic components of a good screening program to determine the relative effectiveness of physiological versus questionnaire components. The results were not surprising: Physiological measures were better predictors of physiological outcomes than questionnaire data. Both groups reported the costs of the physiological data measures to be considerably (approximately two to four times) higher than the questionnaire data. Hall (1980) found that an HRA (questionnaire only) is more cost-effective than physical exams in detecting serious medical problems.

Effects of Incentives in Encouraging Changes in Health Behaviors and Risks in an Employee Population

The use of financial incentives by employers to encourage healthy lifestyle behaviors is a relatively new technique. This strategy is based on the concept that high-risk health behaviors lead to high-cost health care. Studies by Yen et al. (1990, 1991) showed that 10% of employees account for 58% to 80% (multi- versus single-year time periods) of health care costs and that high-risk health behaviors cost more than low-risk health behaviors. For example, annual medical claims (1990 dollars) for smokers were $228 more than for nonsmokers and physically inactive employees cost $391 more than physically active individuals. The authors caution that although individuals with high-risk behaviors are more expensive than low-risk behaviors, it is not yet clear whether decreases in health care costs will follow a reduction in these modifiable risk factors. Edington and Yen (1992a) recommend that health promotion programs be evaluated on their ability to lower employee health risk behaviors and maintain individual well-being and quality of life, rather than on their ability to lower health care costs.

The most common use of incentives is to encourage participation in wellness programs. Madlin (1991) reviewed the use of incentives and recommended that neither mandatory participation nor disincentives, which penalize unhealthy employees, should be used. The most popular incentives and ones with benefits far beyond the actual cost are t-shirts, gym bags, books, hats, and gift certificates.

The use of incentives is not a new technique in the wellness field. George Pfeiffer was very successful in the early days of the Xerox program in conducting a low-cost wellness communications program to over 400 Xerox sites throughout the country. In his program and in others even the cost of a weekend vacation for two to some exotic location significantly increased participation in a cost-effective way.

Caldwell (1992) cites the use of incentives at seven companies and states that 3% of the 235 Fortune 500 industries he studied currently use financial incentives to reward employees for what are viewed as healthy behaviors. Another 9% were expected to introduce incentives by the end of 1992, and another 19% are considering the strategy. Caldwell reports that Southern California Edison employees are eligible for a $10 monthly rebate if they pass five health tests (body weight, blood pressure, cholesterol, blood glucose, and carbon monoxide levels). At Baker-Hughes, Inc., employees pay a $10 per month surcharge for using tobacco and receive a $100 pretax contribution to their health care reimbursement account if they meet the minimum health standards in three tests (cholesterol, blood pressure, and body weight). The company expected to save 3 to 5 million dollars over 3 years as a result of lowered lifestyle behavior risk. This is a very optimistic goal and one that this author would not expect to be achieved. Dominion Resources, Inc. rewards employees with $10 a month if they wear seat belts, don't smoke, and maintain acceptable blood pressures, body weight, and cholesterol levels. Bernstein (1990) cites companies including Adolph-Coors, Southern California Edison, and Ventura County in California for reducing the cost to the employee for health care insurance when the employee exhibits positive health behaviors.

Overman and Thornborg (1992) describe the efforts at The Moore Company, a small electronics firm with 65 employees, which cut its contribution to the employee health plan in half. Moore then allowed employees to earn back those dollars and up to 25% more by reporting healthy lifestyles. The Moore Company's cost of health care decreased from 8.4% to 7.2% of salaries in 2 years. Schott and Wendel (1992) describe the wellness program at Central States Health and Life Company of Omaha, focusing on smoking, alcohol, nutrition, and exercise programs for its employees. The primary goal was to provide a low-cost program for its 585 employees in order to control health insurance premiums without shifting the costs to employees. It is still too early to report any cost savings. Likewise, Santora (1992) describes Sony's program for its 12,000 employees, providing preventive care coverage with full reimbursement for blood screening, pap smears, mammography, and selected reimbursement on other health promotion programs. The wellness programs, implemented by an aggressive provider organization, are expected to decrease the rate of annual increase and thereby save claims cost over a long period of time.

Cave (1992) presents an argument to show that financial incentives or disincentive programs could be expected to save from 1% to 5% of health care claims expenses. He cites examples of companies that pay one-time lump sum incentives for nonsmoking or weight loss. Other examples include preventive care accounts that pay employee enrollment fees in health promotion programs. A third example of incentive programs was increased employer premium contributions. He argues that disincentives are more likely to save more money. Cave stressed the return-on-investment advantages of disincentive programs such as condition-specific deductibles. In this type of program employees are assessed an extra $500 to $1500 deductible for medical expenses for accidents associated with specific lifestyle behavior, such as accidents related to intoxication, illegal drugs, or improper use of seat belts or safety helmets. Other disincentives are coinsurance differentials for using physicians outside of the PPO network and penalizing, for example, smokers with a 10% less coinsurance.

Condition-specific coinsurance and premium contributions are discussed by Caudron (1992a) to illustrate the ways companies are using various forms of incentives and disincentives to encourage healthy lifestyle behaviors. The author describes the pilot program at Hershey Foods, where the monthly assessment for unhealthy lifestyle behaviors can amount to $43 per month for smoking, $40 for high blood pressure, $12 for cholesterol, $16 for lack of exercise, and $36 for excess body weight. Employees who meet all five healthy criteria receive an extra $30 per month off their health care cost premiums, whereas those who have all five risk factors pay an extra $117 per month.

Priester (1992) raises a concern about the fairness of financial incentives, as it is not yet clear which health risk behaviors are totally voluntary. This is a valid concern because incentives are based on the concept that individuals who voluntarily engage in poor health behaviors should not be entitled to the same health care resources as those who live healthier lifestyles.

Economic Impact of Changes in Health Behaviors and Risks

The most interesting and relevant economic question in the disease prevention, health promotion, and wellness movement is, "If individuals change health behaviors and risks, will their associated economic indicators change, and, if so, at what time?" To begin to examine this question, we can look at each of the behavior-change and risk-change programs independently.

Weight Loss

Weight-loss programs are typically among the most popular health behavior programs and yet often the least successful. Cameron, MacDonald, Schlegel, Young, and Fisher (1990) invited community volunteers through local media advertising to participate in a correspondence weight-loss program. The participants were a highly biased group, as only 14 males and 128 females participated in the 15-week program. At the end of 1 year, 13 men reported an average weight loss of nearly 13 pounds and the 128 women reported an average weight loss of approximately 5 pounds. The authors claimed the program cost about $5 for each pound lost. This is a biased number since the authors used only the most "involved" women, limiting their financial analysis to the 96 women who claimed to have read at least 8 of the 15 lessons and attended one of the two interim weigh-ins. This involved group represented 65% of the initial population.

Smoking Cessation

The cost of smoking (which still accounts for one of every six deaths in the United States) has been analyzed more than the cost of any other wellness-related behavior. In 1990, Elixhauser reviewed the cost of smoking cessation programs. Based solely on medical expenditures, the rest-of-lifetime costs of smoking range from $41,000 for a middle-aged smoker to $400 for an elderly female smoker. The cost-effectiveness data determined that the cost per "quitter" was between $20 to $400. These types of data are hard to calculate because previous history of quitting is often not taken into account. It is well documented that the person most likely to succeed in a smoking cessation program is one who has either taken the course before or has tried previously to quit smoking. The studies reviewed by Elixhauser showed a $4,000 to $9,000 cost per life-year saved. For television audience programs versus smoking cessation class programs, the cost per quitter was $27 and $75, respectively. It was suggested that one of the most promising cost-effective methods to reduce smoking is through the use of legal regulations and taxation. These interventions appear to have the highest benefit-to-cost ratios, causing smoking to decline an estimated 4% to 14%.

Marks, Koplan, Hogue, and Dalmat (1990) studied the theoretical benefits of smoking cessation for pregnant women. The authors utilized the 1985-1986 behavior risk factor surveillance system and concluded that offering smoking cessation programs in prenatal care programs could produce a 3.0 to 7.0 benefit-to-cost ratio.

Hypertension

Foote and Erfurt (1991) analyzed 4 years of medical claims data for hypertensive and normotensive subjects. The authors cite a reduction in health care claims for the hypertensive group undergoing treatment. The data are in the right direction, but the authors used inappropriate statistics. The standard deviations of the claims data were several times greater than the means and the data were severely skewed to the right.

Cholesterol

Wilson, Edmundson, and DeJoy (1992) conducted an extensive analysis of the cost-effectiveness of a cholesterol screening program sponsored by a company with 40,000 employees at 400 different locations. The authors selected 37 locations representing nearly 6,000 employees. Approximately 54% of the 6,000 eligible employees participated in the cholesterol screening and intervention programs. The most cost-effective program was the 1-month educational program with an incentive (low-cost prize) for cholesterol reduction. However, the authors demonstrated that the screening-only group was nearly as effective as the more comprehensive education- and incentive-based programs. Participation was the most important factor in determining cost-effectiveness ratios. It is clear that increasing the participation rate maximizes the cost-effectiveness of intervention. The cost of the program was between $8.21 and $13.85 per employee screened (average of $9.75) while the total cost of the study per participant ranged between $27 to $60 (with an average of $37.78 per participant). In addition to the screening, the cost for the interventions and project administration averaged $9.36 per participant (with a range of $6.60 to

$21.24). The authors were able to encourage 28% of the participants initially classified as borderline high or high risk to reduce their cholesterol by 10%. The cost of this reduction was $286 per participant. The full impact of the study in relation to cost-benefit of health promotion programs or risk-benefit of programs is not clear, as no estimate was made of their risk-reduction or benefit to the organization in terms of productivity or other financial returns.

Drug Screening

Zwerling, Ryan, and Orav (1992) described the cost benefit of a pre-employment drug screening program for the Postal Service in Boston, Massachusetts. The authors utilized assumptions of the prevalence of positive tests, the rate of absenteeism of drug users versus non–drug users, and the more frequent rates of accidents, injuries, and turnover for users. The authors also used averages for the Boston area postal workers on wages and benefits, costs of rehabilitation, and turnover for employees who used drugs. It was estimated that drug testing cost approximately $108 per applicant hired. This cost was compared to the savings in absenteeism, accidents, injuries, and turnover avoided of $270 per applicant hired. The analysis is sensitive to prevalence of drug users in the population. The authors also list many other costs and benefits that could be used in the analysis, but overall they perceive a high benefit-to-cost ratio for these programs given that drug user prevalence among workers is more than 2%.

Physical Fitness

Shephard (1992a) found over 15 companies reporting cost savings or utilization reductions related to participation in employee fitness programs. The health care cost savings ranged from $61 per year to $450 per year for participants versus nonparticipants. He also cites 17 companies that reported significant lifestyle behavioral changes related to their participation in the employee fitness program. Shephard estimates the cost of exercise facilities at $150 to $500 per participant per year. His summary of cost-benefit ratios includes 14 studies with $1.07 to $5.78 returned for each dollar invested. He concluded that the programs are still relying on relatively short-term data that may or may not change over longer evaluation time periods. Based on the nature of the worksite, Shephard recommends that companies stay with low-cost programs if the goal is to gain a high yield on investment. The cost analysis presented by Shephard did not take into account process-oriented outcomes and other less quantifiable outcomes such as recruitment, adherence, turnover, productivity, absenteeism, and perceived health or predicted future risk, although Shephard discussed these outcomes as other potential benefits.

Shephard (1992b) reviewed the 12-year data from Canada Life Assurance Company's study of employee fitness. Employees who are members of the fitness center pay $55 a year and the company pays $220 per year per member. Enrollment in the relatively small (5,000 square feet) health club facility averaged approximately 13% over the 12 years. Shephard found that turnover and absenteeism were lower among the participants than nonparticipants. After 12 years, Shephard estimates that the benefits of the program were approximately $340 per worker spread across all employees, while the cost of the program was approximately $450, including facility rental expenses. This is another example where the analysis is apparently sensitive to participation levels.

Medical Clinics

Perhaps an effective way to address employee health behaviors and risks is for large worksites to offer medical clinics. This could help the organization control medical treatment and, to some extent, cost. The cost savings could result from improved direct medical care, less time spent by the employee to find medical care, and in a lower cost of prescription drugs.

Another advantage of an on-site medical unit is its medical surveillance function. Bond, Lipps, Stafford, and Cook (1991) found that workers who did not participate in a company-sponsored program had 23% higher mortality within the first 5-year follow-up. After 15 years there was no difference in the two groups. The implications from this study are not clear. It could be that only the "healthy" workers volunteered for the surveillance program, or it could be that the program initially motivated workers to correct any immediate problems. The potential benefit is for companies to defer health care costs until later in life when Medicare is available to supplement the company's responsibility.

Monitoring Environments for Risks

An evolving critical issue at the worksite is the interaction of the work environment and personal lifestyle behaviors and risks. Walsh, Jennings, Mangione, and Merrigan (1991) surveyed 2,700 industrial workers about their perceived work environment, health risks, and health behaviors. The data form a strong case for combining worksite wellness efforts with occupational medical efforts. Employees with either few job risks or few lifestyle risks were the least likely to report injuries. Those employees with both high lifestyle risks and job risks were most likely to report injuries. As injuries contribute to health care costs, it is clear that both job environmental risks and lifestyle-related risks need to be considered when attempting to control health care costs.

Summary and Suggestions for Further Research

The efficacy of worksite health promotion interventions in reducing health risks has been demonstrated in the recent past and is discussed in other chapters in this volume. The data are often tracked by worksites and often compared to community and national data, especially the Healthy People 2000 project.

The risk-cost relationship is the rationale for "risk-rating" health insurance. This strategy is usually based on one or more modifiable risk factors. Typical threshold values used to categorize high risks include: cholesterol (240 mg/100 ml of blood or above); blood pressure (90 mm Hg diastolic and 140 mm Hg systolic); body weight (120% or 130% of the 1958 Metropolitan Life Tables); safety belt use (74% use or less); vigorous physical activity (less than once per week); use of alcohol (21 drinks per week or more); and use of cigarettes (current smoker). Most high-risk individuals have high health care costs. The excess health care costs associated with each high-risk behavior ranges from $150 to $400 per year (1990 dollars). The excess costs are not additive but there is evidence that increasing numbers of high-risk classifications are associated with up to $1,300 per year excess costs for an individual with six or more high-risk classifications compared to an individual with zero high risks.

The most pressing issue related to the use of health behaviors and risks assessment in risk-rating is whether reduced risks link to future lower costs. If they are linked, within what time frame? Some of the data presented in the third section of this chapter support the linkage, but it is still too early to draw conclusions. Until this issue is resolved, health promotion should continue to emphasize the educational and motivational value of the assessment of health behaviors and risks rather than focusing on its cost implications.

Research Priorities

Following are several questions to be explored in further research:

1. To what degree can excess health care costs be attributed to risks?
2. What are the age- and gender-specific criteria for risks and costs?
3. How can risks and costs assessments be improved?
4. Do "risk-rating" health insurance practices reduce costs and risks?
5. Do changes in health risks and behaviors have other economic implications?

Tracking the Data

The most promising recent data management tool is a system that merges all the relevant data bases and asks interrelated questions. The integrated health data management system, as described by Yenney (1990, 1992) and Golaszewski, Kaelin, Miller, and Douma (1992), has the potential to answer many health- and cost-related questions. However, it is still the responsibility of managers to formulate the most important questions and then to devise effective solutions. Data bases used by the University of Michigan Fitness Research Center (1993) include health insurance claims (utilization and costs by ICD-9 classifications), health risk appraisal data, personnel data, productivity measures, employee assistance program utilization, medical department records, workers' compensation claims, disability claims, and health promotion program participation. As Yenney points out, there are long-term commitment problems, turf battles, and confidentiality issues that individual companies must address. This integrated health data management system is probably not viable for every company but yields insight into generic problems which, when solved, will provide general information for the total wellness field. Some of the early data from these types of analyses have led observers to associate excess medical care costs with negative health behaviors and health risks. The conclusions have, in turn, led to the practice of tying incentives and disincentives to health behaviors and risks. A more positive objective will be the use of these data systems to improve the synergy and effectiveness of benefit programs, health promotion and wellness programs, medical services, safety, and other human resource programs.

References

Bernstein, A. (1990, May). Health care costs: Trying to cool the fever. *Business Week*, pp. 46-47.

Bertera, R.L. (1991). The effects of behavioral risks on absenteeism and health-care costs in the workplace. *Journal of Occupational Medicine*, **33**, 1119-1124.

Bond, G., Lipps, T.E., Stafford, B.A., & Cook, R.R. (1991). A comparison of cause specific mortality among participants and non-participants in a worksite medical surveillance program. *Journal of Occupational Medicine*, **33**, 677-680.

Brink, S.D. (1986). Health risks and behaviors: The impact on medical costs. A preliminary study by Millman and Robertson Inc. and Control Data Corporation.

Caldwell, B. (1992). Companies implement wellness incentives to help control health plan costs. *Employee Benefit Plan Review*, **46**(11), 50-52.

Cameron, R., MacDonald, M.A., Schlegel, R.P., Young, C.I., & Fisher, S.E. (1990). Toward the development of self-help health behavior change programs. Weight loss by correspondence. *Canadian Journal of Public Health*, **81**, 275-279.

Caudron, S. (1992a). Are health incentives disincentives? *Personal Journal,* **71**(8), 35-40.

Caudron, S. (1992b). A low cost wellness program. *Personal Journal,* **71**(2), 34-38.

Cave, D.G. (1992). Employees are paying for poor health habits. *Human Resource Magazine,* **37**(8), 52-58.

Conrad, K.M., Conrad, K.J., & Walcott-McQuigg, J. (1991). Threats to internal validity in worksite health promotion program research: common problems and possible solutions. *American Journal of Health Promotion,* **6**(2), 112-122.

Dewey, J.E., & Seehafer, R.W. (1987). Health risk appraisal (HRA): An accurate assessment of a company's health status and health risk. In *Proceedings of the 23rd annual meeting of the Society of Prospective Medicine* (pp. 125-132), Atlanta.

Edington, D.W., & Yen, L.T-C. (1992a). It is possible to simultaneously reduce risk factors and excess health care costs? *American Journal of Health Promotion,* **6**(6), 403-406, 409.

Edington, D.W., & Yen, L.T-C. (1992b). The validity and reliability of HRAs. In K.W. Peterson and S.B. Hilles (Eds.), *The Society of Prospective Medicine directory of health risk appraisals.* Indianapolis, IN: Society of Prospective Medicine.

Elixhauser, A. (1990). The costs of smoking and the cost effectiveness of smoking cessation programs. *Journal of Public Health Policy,* **11**(2), 218-237.

Fitness Research Center (1993). *Integrated health data management system.* University of Michigan.

Foote, A., & Erfurt, J.C. (1991). The benefit to cost ratio of worksite blood pressure control programs. *Journal of the American Medical Association,* **265**(10), 1283-1286.

Foote, A., & Heirich, M.A. (1991). Screening vs. appraisals: Which is most effective in involving employees in improving health? *Employee Assistance,* **3**(6), 22-26.

Foxman, B., & Edington, D.W. (1987). The accuracy of health risk appraisal in predicting mortality. *American Journal of Public Health,* **77**(8), 971-974.

Froom, J., & Froom, P. (1990). Consequences of the National Cholesterol Education Program. *Journal of Family Practice,* **30**(5), 533-536.

Garfinkel, S.A., Riley, G.F., & Iannacchione, V.G. (1988). High cost users of medical care. *Health Care Financing Review,* **9**(4), 41-52.

Gazmararian, J.A., Foxman, B.F., Yen, L.T-C., Morgenstern, H., & Edington, D.W. (1991). Comparing the predictive accuracy of health risk appraisal: The Centers for Disease Control versus Carter Center Program. *American Journal of Public Health,* **81**(10), 1296-1301.

Golaszewski, T., Kaelin, M., Miller, R., & Douma, A. (1992). Integrated health care cost management strategy. *Benefits Quarterly,* **8**(1), 41-50.

Hall, J.H. (1980). Which health screening techniques are cost effective? *Diagnosis*, (February), 60-82.

Holder, D., & Blose, J.O. (1991). A comparison of occupational and non-occupational disability payments and work absences for alcoholics and non-alcoholics. *Journal of Occupational Medicine*, **33**(4), 453-457.

Holder, D., & Blose, J.O. (1992). The reduction of health care costs associated with alcoholism treatment: A 14-year longitudinal study. *Journal of Studies on Alcohol*, **53**(4), 293-302.

Imrey, H., Williams, B., Schmale, J., Imrey, P., & Moll, J. (1983). Risk factors and breast cancer screening recommendations: A case control. In *Proceedings of the 18th annual meeting of the Society of Prospective Medicine* (pp. 405-412). Ottawa, ON.

Kannel, W.B., & McGee, D.L. (1987). Composite scoring-methods and predictive validity: Insights from the Framingham study. *Health Services Research*, **22**(4), 499-535.

Kramer, D., Wiley, J., & Camacho, T. (1981). A study of the predictive validity of the Health Hazard Appraisal. Working Paper, California Department of Health Services, Berkeley, CA.

Lu, C-F.C., Yen, L.T-C., & Edington, D.W. (1993). *The relationship between health behavioral risks and health care costs and utilization due to cardiovascular disease in a manufacturing company*. (Tech. Rep. to the Michigan Health Care and Research Foundation). Detroit.

Lynch, D., Teitelbaum, H.S., & Main, D.S. (1991). The inadequacy of using means to compare medical costs of smokers and nonsmokers. *American Journal of Health Promotion*, **6**(3), 206-213.

Madlin, N. (1991). Wellness incentives: How well do they work? *Business and Health*, **9**(9), 70-74.

Marini, G.A. (1991). Comprehensive drug-abuse program can prove effective in the workplace. *Occupational Health and Safety*, **60**(4), 54-59.

Marks, J.S., Koplan, J.P., Hogue, C.J., & Dalmat, M.E. (1990). A cost-benefit/cost-effectiveness analysis of smoking cessation for pregnant women. *American Journal of Preventive Medicine*, **6**(5), 282-289.

Muchnick-Baku, S., & McNeil, C. (1991). *Healthy People 2000 at work: Strategies for employers*. Washington, DC: The National Resource Center on Health Promotion, Washington Business Group on Health.

National Center for Health Services Research (1987). A research agenda for personal health risk assessment methods in health hazard/health risk appraisal. *Health Services Research*, **22**(4), 441-622.

Office on Smoking and Health (1989). *Reducing the health consequences of smoking: 25 years of progress*. A report of the Surgeon General (DHHS Publication No. ([CDC] 89-8411). Washington, DC: U.S. Department of Health and Human Services.

Overman, S., & Thornburg, L. (1992). Beating the odds. *Human Resource Magazine*, **37**(3Z), 42-47.

Pavett, C., & Whitney, G.G. (1990). Exercise makes employees work better. *Human Resource Management*, **35**(12), 81-84.
Penner, S., & Penner, S. (1992a). Excess insured health care costs from tobacco-using employees in a large group plan. *Journal of Occupational Medicine*, **32**(6), 521-523.
Penner, S., & Penner, S. (1992b). Hospital utilization of tobacco users vs. nonusers in an HMO. *Health Values: Achieving High Level Wellness*, **16**(1), 17-22.
Priester, R. (1992). Are financial incentives for wellness fair? *Employee Benefits Journal*, **17**(1), 38-40.
Robbins, L.C., & Hall, J. (1970). *How to practice prospective medicine*. Indianapolis, IN: Methodist Hospital of Indianapolis.
Santora, J.E. (1992). Sony promotes wellness to stabilize health care costs. *Personnel Journal*, **71**(9), 40-44.
Schott, W., & Wendel, S. (1992). Wellness with a track record. *Personnel Journal*, **71**(4), 98-104.
Shephard, R.J. (1992a). A critical analysis of worksite fitness programs and their postulated economic benefits. *Medicine and Science in Sport and Exercise*, **24**(3), 354-370.
Shephard, R.J. (1992b). Twelve years' experience of a fitness program for the salaried employees of a Toronto Life Assurance Company. *American Journal of Health Promotion*, **6**(4), 292-301.
Shipp, M., Croughan-Minihare, M.S., Pettit, D.B., & Washington, A.E. (1992). Estimation of the break-even point for smoking cessation programs in pregnancy. *American Journal of Public Health*, **82**(3), 383-390.
Siegelman, S. (1991). Employers fighting the battle of the bulge. *Business and Health*, **9**(12), 62-73.
Suggs, T.F., Cable, T.A., & Rothenberger, L.A. (1990). Results of a worksite educational and screening program for hypertension and cancer. *Journal of Occupational Medicine*, **32**(3), 220-225.
Szymanski, L., Pate, R.R., Dowda, M., Blair, S.N., & Howe, H.G. (1991). A comparison of questionnaire and physiological data in predicting future chronic disease risk factor status in an employee population. *American Journal of Health Promotion*, **5**(4), 298-304.
U.S. Department of Health, Education and Welfare (1979). *Healthy people: The Surgeon General's report on health promotion and disease prevention* (DHEW [PHS] No. PHS 79-55071). Washington, DC: U.S. Government Printing Office.
U.S. Department of Health and Human Services (1980). *Promoting health/preventing disease: Objectives for the nation*. Washington, DC: U.S. Government Printing Office.
U.S. Department of Health and Human Services (1990). *Healthy people 2000: National health promotion and disease prevention objectives*. Washington, DC: U.S. Government Printing Office.

U.S. Department of Health and Human Services (1993). Cigarette smoking among adults—United States, 1991. *Morbidity and Mortality Weekly Report*, **42**(12), 230-233.

Walsh, D., Jennings, S.E., Mangione, T., & Merrigan, D.M. (1991). Health promotion versus health protection? Employees' perceptions and concerns. *Journal of Public Health Policy*, **12**(2), 148-164.

Wiley, J.A. (1980). Predictive risk factors do predict life events. In *Proceedings of the 16th annual meeting of the Society of Prospective Medicine* (pp. 75-79). Tucson, AZ.

Wilson, M.G., Edmundson, J., & DeJoy, D.M. (1992). Cost effectiveness of worksite cholesterol screening intervention programs. *Journal of Occupational Medicine*, **34**(6), 642-649.

Yen, L.T-C. (1990). *The economic impact of employee health on the organization*. Unpublished doctoral dissertation, Division of Kinesiology, University of Michigan.

Yen, L.T-C., & Edington, D.W. (1988). Using HRA to predict health care costs. In *Proceedings of the 24th annual meeting of the Society of Prospective Medicine* (pp. 251-256). St. Petersburg, FL.

Yen, L.T-C., Edington, D.W., & Witting, P. (1991). Associations between health risk appraisal scores and employee medical claims costs. *American Journal of Health Promotion*, **6**(1), 46-54.

Yen, L.T-C., Edington, D.W., & Witting, P. (1992). Prediction of prospective medical claims and absenteeism costs for 1284 hourly workers from a manufacturing company. *Journal of Occupational Medicine*, **34**(4), 428-435.

Yenney, S.L. (1990). Solving the health data management puzzle. *Business and Health*, **8**(9), 41-49.

Yenney, S.L. (Ed.) (1992). *Putting the pieces together: A guide to the implementation of integrated health data management systems*. Washington, DC: Washington Business Group on Health.

Zwerling, C., Ryan, J., & Orav, J. (1992). Costs and benefits of pre-employment drug screening. *Journal of the American Medical Association*, **267**(1), 91-93.

Chapter 7

Worksite Health Promotion and Injury

David Chenoweth

In the past decade rising health care costs have gained nationwide attention, due in part to their impact on corporate profits. Many factors are driving today's health cost spiral—inflation, new technology, greater utilization, and cost-shifting, to name a few. While the bulk of a company's health care costs may be tied to heart disease and other catastrophic diseases and illnesses, the rising cost of occupational injuries is also a financial strain on many U.S. employers.

This chapter presents an overview of the incidence, types, and costs of major occupational injuries confronting many employers and various health promotion strategies used to decrease the impact of such injuries.

Injury Incidence, Types, and Costs

In 1970 the Occupational Safety and Health Act was established to encourage and mandate American employers to provide safe workplaces for their employees. Since then the incidence of occupational illnesses and injuries has decreased nearly 24%, yet the number of lost workdays has increased nearly 62%, as illustrated in Figure 7.1 (U.S. Department of Labor, 1992). Despite the decreased incidence, employers reported about 6.3 million occupational injuries and illnesses in 1991, with an injury to illness ratio of 20:1 (U.S. Department of Labor, 1992).

According to the National Institute of Occupational Safety and Health (1993), the 10 most common occupational injuries and illnesses are

1. cumulative trauma disorders,
2. occupational lung disorders,

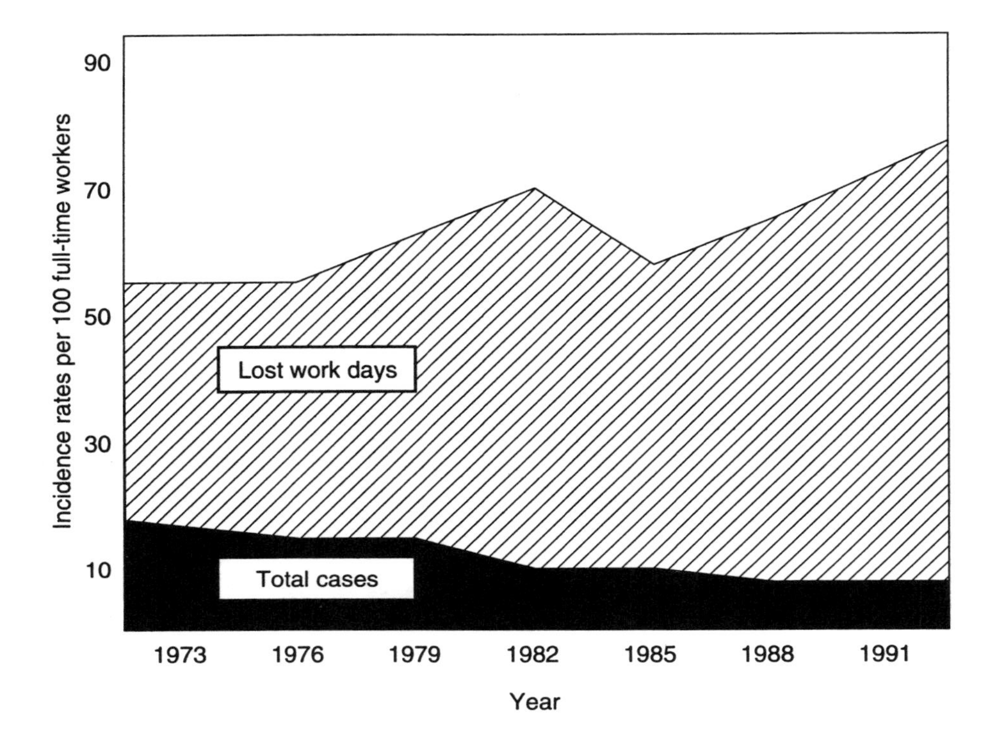

Figure 7.1 Occupational injury and illness incidence rates compared to lost work days, 1973-1991.
Note. From Newsletter of the U.S. Department of Labor, Bureau of Labor Statistics (pp. 1 and 11), November 18, 1992.

3. occupational cancers (other than lung),
4. severe occupational traumatic injuries,
5. cardiovascular disorders,
6. disorders of reproduction,
7. neurotoxic disorders,
8. noise-induced loss of hearing,
9. dermatologic conditions, and
10. psychological disorders.

Nearly half (2.9 million) of the cases reported in 1991 were serious enough to require workers to have their work activity restricted or to lose worktime (U.S. Department of Labor, 1992).

Cumulative trauma disorders (CTDs) are the most common type of occupational injury in the United States (U.S. Department of Labor, 1993). Specifically, about 2 million workers suffer some degree of carpal tunnel syndrome; 4 million workers complain of tendonitis; and about 8.5 million workers suffer CTD-related low back pain (National Center for Health Statistics research department, personal communication, 1993). The following facts reflect this growing crisis:

- Approximately 18 million U.S. workers suffer CTDs annually and half of all workers run the risk of contracting a CTD (*CTD News*, 1992).
- The incidence of CTDs has nearly doubled in the last 3 years, and has climbed almost 600% since 1980. CTDs now account for more than half of all workplace illnesses, as illustrated in Figure 7.2.
- Recent studies indicate that CTDs affect 50% of supermarket cashiers, 41% of meatpackers, 40% of newspaper workers, and 22% of telecommunications workers (NIOSH, 1993).
- Sprains and strains of the torso—specifically in the lower back—account for 50% of all workers' compensation claims and 25% of all health care claims (*CTD News*, 1992). In fact, business executives consider back pain one of their company's top four health problems (*Business & Health*, 1990).
- Approximately 75% of all American workers will experience a back injury—short-term, in most cases—sometime in their working lives (*CTD News*, 1992).

Despite the fact that the majority of CTD treatments are performed in outpatient health care settings, CTD costs have quadrupled since 1987 (Liberty Mutual Insurance, 1992). For example, the average case of carpal tunnel syndrome has risen to $29,000 in workers' compensation expenses and the average case of back pain, the most common CTD, is approximately $24,000 (Fefer, 1992).

What causes CTDs? Various factors may contribute, including poorly designed equipment, fast-paced work, no rest breaks, job and personal stress, poor posture, force and repetition, and individual physical traits (*CTD News*, 1992). Debate continues over whether the individual or the workplace should be the focus of CTD prevention efforts. One study has added to the debate regarding a specific CTD, carpal tunnel syndrome (CTS), by suggesting that obesity is the single most important risk factor for the development of CTS symptoms (Nathan, Keniston, Myers, & Meadows, 1992). Yet, some researchers view these findings with skepticism and strongly contend that a combination of individual and occupational factors are largely responsible for CTS and other CTDs.

Another study investigated personal and job characteristics of musculoskeletal injuries in a large industrial work force for a 2-year period (Tsai, Gilstrap, Cowles, Waddell, & Ross, 1992). Two hundred and seventy-five employees with low back injuries and 456 employees with other types of injuries were compared with over 8,000 employees who did not have a musculoskeletal injury during this period. Injury data were classified only

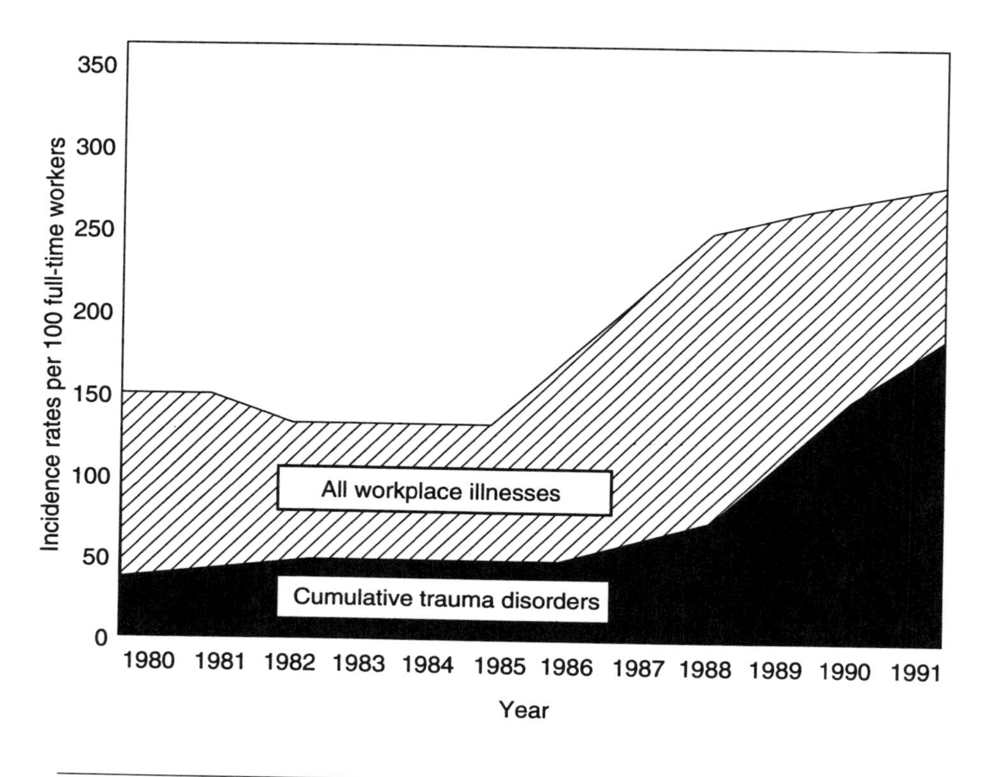

Figure 7.2 Incidence rates for CTDs compared to all workplace illnesses. *Note.* From U.S. Department of Labor Statistics, 1993.

by ICD-9 code to distinguish between the low back and non–low back injuries. In this particular study, both low back and non–low back injuries refer to traumatic events (including both acute and chronic repetitive disorders) serious enough to cause a work absence for more than 5 days. Risk factor data were derived from the employer's Health Surveillance System, which contains all employee preplacement and periodic physical examinations. Multivariate logistic regression analysis revealed statistically significant independent effects of age, obesity, and smoking for both low back and non–low back injuries. Moreover, persons with potentially more physically demanding jobs had higher injury rates in both categories, especially for low back injuries. Despite the favorable outcomes, two possible limitations should be considered in their interpretation. First, job title was used as a surrogate for the physical demands a person may actually experience at work. Kelsey and Golden (1987) described this issue and the possibility of job exposure misclassification—that is, persons who already had low back pain may have changed jobs. Such a misclassification can affect the strength of the association between physically demanding jobs and low back pain. Second, there were no data provided on each subject's physical fitness level, despite some evidence that physically fit persons incur fewer and less severe musculoskeletal ailments than unfit persons (Biering-Sørensen, 1984; Chaffin, Herrin, & Keyserling, 1978; Whitmer, 1992).

Although the previous studies have some limitations, they generate some important questions, including: What types of personal factors actually influence injury risk? Are these factors as influential at various types of worksites or only in a specific type of worksite? Can specific factors be influenced by voluntary lifestyle changes resulting from a worksite health promotion intervention? Which intervention makes the greatest impact for the least cost? If a worksite health promotion intervention carries great potential for success, are American employers ready to and capable of implementing it?

Healthy People 2000

In an attempt to further promote the health status of all Americans, the Office of Disease Prevention and Health Promotion and the U.S. Department of Health and Human Services developed *Healthy People 2000*, a listing of 300 objectives for organizations to use as a guideline to achieve specific health promotion targets by the end of this decade (National Resource Center on Worksite Health Promotion, 1991). Approximately 25 objectives are specific to worksites and relate to the broad challenge of injury prevention. For example:

- **Objective 10.2**: Reduce work-related injuries resulting in medical treatment, lost time from work, or restricted work activity to no more than 6 cases per 100 full-time workers. (Baseline: 7.7 cases per 100 in 1987.)
- **Objective 10.3**: Reduce cumulative trauma disorders to an incidence of no more than 60 cases per 100,000 full-time workers. (Baseline: 100 cases per 100,000 in 1987.)
- **Objective 10.13**: Increase to at least 50 percent the proportion of worksites with 50 or more employees that offer low back injury prevention and rehabilitation programs. (Baseline: 28.6 percent offered back care activities in 1985.) (p. 105)

Worksite Injury Prevention Efforts

Some health care professionals suggest that workers who regularly engage in vigorous, "whole body" exercise such as swimming or jogging have a much lower risk for carpal tunnel syndrome (CTS) and, possibly, other types of CTDs (*CTD News*, May 1992). Several published studies have investigated the possible relationship between worksite health promotion and occupational injuries with the greatest emphasis on low back injuries. In about half of these studies benefits and costs are measured within a benefit-cost analysis (BCA). Benefits are typically quantified in terms of actual or estimated averted costs attributed to one or more of the following categories:

- Lost time (injury-related absenteeism)
- Medical care costs associated with injuries
- Workers' compensation costs associated with injuries

In the remaining articles, researchers use more normative evaluations by simply comparing the number of back injuries before and after a particular intervention, without attaching a monetary cost per injury.

In response to the growing number of CTDs, employers have expanded injury prevention efforts to include

- specific physical fitness programs,
- prework stretching routines,
- flexibility enhancement exercises, and
- multifaceted approaches.

Physical Fitness Programs

Although numerous articles support worksite physical fitness programs (Shephard, 1992), little attention has been given to study the relationship between physical fitness and occupational injuries. The literature on injury risk associated with exercise consists largely of clinical case reports of athletes engaged in intense, competitive exercise (Koplan et al., 1982, 1985; Walter, Sutton, & McIntosh, 1985). Information on the risk and cost of injuries among participants in employee fitness programs or adult recreation-type exercise is also extremely limited (Felton, 1971; Shephard, 1986).

Tsai, Bernacki, and Baun (1988) studied the prevalence, cost, and type of injury among participants of an employee fitness program and nonexercising co-workers over a 2-year period. Overall, there were no significant differences in the rate or cost of injuries among the various participant levels (from 0 to 3 or more times per week). However, the data indicated that individuals who occasionally participated in the fitness program experienced a greater, but nonsignificant, risk of injury (6.3 injuries per 100 persons who exercise less than once a week, and 7.7 injuries per 100 persons who exercise 1 to 2 times a week) than nonparticipants (5.7 injuries per 100 persons). However, injury prevalence was lowest among persons who exercised 3 or more times a week (5.4 injuries per 100 persons) as was the resultant per capita cost of injuries ($32 for regular participants vs. $42 for nonparticipants). Moreover, injury prevalence dropped most notably in persons 50 years of age or older—the more sessions the greater the drop. The authors concluded that the impact of exercise at an onsite fitness facility on overall injury rates and costs among employees is negligible.

In another study, the impact of a physical fitness program on major medical and disability costs was investigated in a large insurance company (Bowne et al., 1984). Employees were involved in a worksite fitness program that included a physical examination, periodic screenings, and various health promotion activities. The participation rate was 19% and baseline data for the year prior to the study were compared to those for the year after entering the program. Results showed that the higher the level of employees' fitness, the lower the major medical and disability costs. Counter to the trend for all employees of higher payments over the previous 4 years, there was a 45.7% drop in major medical costs and a 31.7% drop in disability costs, resulting in an estimated benefit-to-cost ratio of $1.93 to $1.00.

One of the strengths of this study was its prospective, longitudinal evaluation of the effects of a health promotion program on two measured cost outcomes. However, the study had several shortcomings that undermine its generalizability to other settings, such as:

- selection bias (the probability that the participants were not representative of the employee population),
- a noncomparable control group,
- no monitoring of fitness program adherence, and
- costs for laboratory tests, physical examinations, and equipment purchases were excluded from the benefit-cost analysis.

A study conducted by Canadian researchers investigated the impact of a physical fitness program on job-related injuries and associated costs at a municipal worksite (Shore, Prasad, and Zroback, 1989). Each of the 134 participants was tested for back fitness, strength, aerobic power, flexibility, weight, body fat percentage, blood pressure, lifestyle, and productivity. Participants were given an exercise prescription based on their overall fitness level. After 6 months of exercising, participants were retested and exhibited a 14.2% increase in their overall back fitness level. Moreover, injury-related absences dropped one-fourth of a day while nonparticipants' absences increased approximately 3.1 days, producing an estimated cost-savings of $62,922.

In a related study involving municipal ambulance drivers and attendants, lost-time accidents dropped 44% among physically fit employees in addition to the number of lost work shifts and workers' compensation costs (Shore et al., 1989).

The preceding results are quite similar to the investigation by Cady, Bischoff, O'Connell, et al. (1979) to determine the effect of physical fitness on back injuries in fire fighters. To ensure high fitness levels, the program design provided each participant with 3 hr of exercise per week and periodically assessed them for the duration of the study. Overall, the study indicated that physical fitness and conditioning actually prevented back injuries. Moreover, the study showed a statistically significant drop in the number of injuries sustained with a corresponding gain in physical fitness.

A follow-up study conducted by Cady (1985) spanned 14 years and showed that enhanced fitness levels strongly corresponded to lower injury rates and associated costs. The physical fitness program was available to all employees on a voluntary basis and a series of physical fitness measurements were done on participants and nonparticipants at various intervals. Although the actual percentage of participants was not reported, it appears that the long-term fitness program contributed to a departmental decrease in total injuries, disabling injuries, and cost per injury. Specifically, Cady found that the fittest employees had only one eighth as many injuries as the least fit employees and that unfit workers incurred twice as many low back injury costs as fit workers. In addition, workers' compensation claims dropped by half for the entire department in the last 8 years of the study and disability costs declined 25%. Nonetheless, it should be noted that these improvements are probably the result of both the physical fitness program and changes in the administration of the return-to-work program.

Thompson (1990) studied the effect of two 5-min exercise breaks on musculoskeletal strain among data-entry operators. The exercises were designed to relieve cervicobrachial posture strain but also included arm, wrist, and lower leg manipulation. In the 2 years before the introduction of the exercise program, there were, at any one time, 7 to 12 active CTD-based workers' compensation claims on behalf of injured operators. In the year following the introduction of the program, there were no new claims and all of the operators out on disability leave returned to work or left the company for personal reasons. There was also an immediate 25% climb in productivity and a cost savings in overtime reported in other areas.

Gates (1988) reported that a back care program consisting of various sitting, standing, and flexibility exercises virtually eliminated back injuries among staff nurses at a Philadelphia hospital. Yet, no time frame for the program was given. The program consisted of 40-min inservice sessions to teach small groups of nurses how to reduce back pain with additional information on nutrition, posture, body mechanics, stress management, and exercise.

What impact, if any, do specific worksite health promotion strategies have on existing musculoskeletal symptoms and injuries? Silverstein and Armstrong (1988) studied the effectiveness of an on-the-job exercise program in reducing the severity of existing musculoskeletal symptoms of discomfort and preventing future occurrences among active workers in a health products facility. Of the original 178 participants, about 80% completed the program. At the conclusion of the program, 67% indicated they felt better, 32% indicated no change, and 1% said the exercising made them feel worse. After 1 year of the program there were no statistically significant differences in either pooled discomfort scores or in the proportion of those whose discomfort improved or worsened between initial and final surveys, based on exercise program participation.

Sirles, Brown, and Hilyer (1991) investigated the impact of back-school education and exercise in back-injured municipal workers. A portion of the sample of 74 employees with back injuries also was assigned randomly to a counseling intervention. Preintervention and postintervention testing revealed significant posttest improvements in back strength and flexibility. Significant improvements were also noted in psychological well-being, depression, anxiety, and perceptions of pain. However, no significant differences were found on any of the measures between employees who did and employees who did not receive the counseling intervention. The authors speculate that methodological factors such as possible sampling bias and instrumentation could have affected the outcomes.

Flexibility Enhancement Exercises

Hilyer, Brown, Sirles, and Peoples (1990) conducted a study involving fire fighters to examine the effect of 6 months of flexibility training on the incidence and severity of joint injuries. Both flexibility and costs (lost time and medical care costs) were studied in a group of 469 fire fighters for 2 years beyond the intervention. Flexibility scores of the experimental subjects were significantly better than the control group. Although the incidence of joint injuries was not significantly different between the experimental group (48 injuries) and the control group (52 injuries), injuries sustained by the former group resulted in significantly less lost-time costs. For example, the group of participants

- used less than half as many health care dollars as nonparticipants ($39,775 vs. $87,550),
- incurred about a third as many lost-time expenses as nonparticipants ($45,597 vs. $147,581), and
- had a lower per-injury cost than nonparticipants ($1,778 vs. $4,522).

In essence, the study showed that specific exercise routines involving shoulder flexion, shoulder extension, sit and reach, and knee flexion contributed to overall greater flexibility.

Stanforth and Plattor (1993) reported that a similar approach in a manufacturing worksite decreased the percentage of employees with musculoskeletal pain from 5% to 1% within a 1-year period.

Landgreen (1990) reported that a 10-minute-a-day stretching and back strengthening program for Coca-Cola's plant employees decreased the accident rate by 83% and produced a yearly savings of over $250 per employee in lost time and replacement costs. The company is in the process of determining whether its dramatic drop in injuries is due to the exercises or to the workers' heightened awareness. Preliminary results suggest it is a combination of the two.

The city of Santa Monica, California, requires employees to participate in a prework back-stretching and strengthening session (Smith, 1990). Early reports show that the program has reduced low back strains and sprains, despite no indication of possible cost-savings.

Multifaceted Approaches to Worksite Injury Prevention

Some researchers contend that most occupational injuries are caused by a combination of environmental (worksite) and personal factors and thus question whether it is really appropriate to expect a single intervention to effectively reduce the risk of such injuries (Mallory & Bradford, 1989). Apparently some companies agree with this notion and provide multiple interventions in an attempt to prevent worksite injuries. For example, Fitzler and Bergner (1983) studied the impact of a multifaceted low back program on the incidence of low back injuries and costs. The program consisted of three components: (1) prevention—low back education and awareness activities, (2) intervention—reporting low back pain and injury immediately, and (3) treatment protocols, ranging from heat applications to mild analgesics. Considering the program's emphasis on encouraging employees to report low back injuries immediately, the authors were not surprised to see the number of low back injuries (LBIs) increase from 20 to 34 in the first 24 months of the program. Yet the number of lost-time LBIs dropped from 10 to 6 injuries and the frequency rate dropped nearly 50% from 2.78 in 1980 to 1.56 injuries per 100 employees in 1982. Moreover, workers' compensation costs for LBIs dropped tenfold from $200,000 to approximately $14,000 in the same time frame.

Morris (1984) reported that a worksite back program based primarily on low back awareness and proper lifting techniques cut low back injuries 95% in the first 2 years of its operation. And Berry (1981) reported that low back injury related absences dropped from 400 days to 19 days at one Burlington Industries worksite due to the implementation of a healthy back program.

One example of a successful worksite back program is the Safeway Corporation program (Chenoweth, 1993). The company took the "worst first" approach (that is, directing resources to the most common and most expensive health risks and claims) and developed a back injury control program with the help of Dr. Arthur White, a noted San Francisco spine surgeon. Program components include videos, posters, hats, belt buckles, newsletters, and competitions. The program teaches people not only how to live a healthy lifestyle but also how to become more physically fit at work. Initiated in 1978 and frequently updated, the program has produced some outstanding results as evidenced by an 18% drop in injuries within the first year. Today, back injuries are 68% below the base year, measured on hours worked.

Finally, Versloot, Rozeman, van Son, and van Akkerveeken (1992) conducted a cost-effectiveness analysis of a "back school" in a Dutch bus company. Objective data on absenteeism were collected and compared during a 6-year period for the control and experimental groups. The longitudinal study revealed that a tailor-made back program reduced back injury related absenteeism by at least 5 days per year per employee and thus was extremely cost-effective to the industry. Moreover, the positive impact was sustained for 2 full years after the program.

Summary and Recommendations

Overall, most of the articles reviewed indicate that specific types of worksite health promotion efforts can reduce the incidence, severity, and associated costs of CTDs. Also, the literature indicates that regular exercisers have no greater risk of sustaining musculoskeletal injuries than nonexercisers and occasional exercisers. Nevertheless, most of the CTD-based studies focus primarily on back injuries and do not address other common occupational ailments such as carpal tunnel syndrome, repetitive motion injury, and mental distress.

Virtually all of the studies reviewed have some methodological shortcomings that limit their generalizability to other worksites. Thus, to minimize these potential drawbacks, future studies should address the following questions:

1. What risk factors predict workers with high injury rates? Do specific risk factors correspond with specific types of injuries?
2. How do the rate and type of injuries on the job compare with injuries incurred during fitness program participation?
3. How can injury data be used to determine the most appropriate injury prevention interventions?
4. Which health promotion interventions are most effective in preventing injuries? Which are most effective in facilitating a speedy recovery and a return to unrestricted work?
5. Are there differences in methods of preventing injuries between small and large worksites?
6. What is the effect of specific health promotion interventions on workers' compensation-based injuries?

References

Berry, C.A. (1981). An approach to good health for employees and reduced health care costs for industry. Health Insurance Association of America, p. 9.

Biering-Sørensen, F. (1984). Physical measurements as risk indicators for low-back trouble over a one-year period. *Spine*, **9**, 106-119.

Bowne, D.W., Russell, M.L., Morgan, J.L., Opterberg, S.A., & Clark, A.E. (1984). Reduced disability in an industrial fitness program. *Journal of Occupational Medicine*, **26**, 809-816.

Business & Health (1990, April). The 1990 national executive poll on health care costs and benefits, p. 37.

Cady, L.D. (1985). Programs for increasing health and physical fitness of firefighters. *Journal of Occupational Medicine*, **27**, 110-114.

Cady, L.D., Bischoff, D.P., O'Connell, E.R., et al. (1979). Strength and fitness and subsequent back injuries in firefighters. *Journal of Occupational Medicine*, **21**, 269-272.

Chaffin, D.B., Herrin, G.D., & Keyserling, W.M. (1978). Preemployment strength testing—an updated position. *Journal of Occupational Medicine*, **20**, 403-408.

Chenoweth, D. (1993). *Health care cost management: Strategies for employers* (2nd ed.). Dubuque, IA: Brown & Benchmark.

CTD News (1992, May). What causes CTS?, p. 1.

CTD News (1992). Abstract in AFB Action Newsletter, Winter 1992, p. 5.

Fefer, M.D. (1992). What to do about workers' comp. *Fortune*, June 27, p. 81.

Felton, J.S. (1971). Occupation. In O.E. Larsen (Ed.), *Encyclopedia of sport sciences and medicine* (pp. 965-974). Macmillan.

Fitzler, S.L. & Berger, R.A. (1983). Chelsea back program: One year later. *Occupational Health & Safety*, **52**, 52-54.

Gates, S.J. (1988). On-the-job back exercises. *American Journal of Nursing*, **88**, 656-659.

Hilyer, J.C., Brown, K.C., Sirles, A.T., & Peoples, L.A. (1990). A flexibility intervention to reduce the incidence and severity of joint injuries among municipal fire fighters. *Journal of Occupational Medicine*, **32**, 631-638.

Kelsey, J.L., & Golden, A.L. (1987). Occupational and workplace factors associated with low back pain. *Occupational Medicine*, **2**, 7-16.

Koplan, J.P., Powell, K.E., Sikes, R.K., Shirley, R.W., & Campbell, C.C. (1982). An epidemiologic study of the benefits and risks of running. *Journal of the American Medical Association*, **248**, 3118-3121.

Koplan, J.P., Siscovick, D.S., & Goldbaum, G.M. (1985). The risks of exercise: A public health view of injuries and hazards. *Public Health Reports*, **100**, 189-195.

Landgreen, M. (1990). Coca-Cola's back care program decreases accident rate by 83%. *Club Industry*, **6**, 15.

Liberty Mutual Insurance. In *CTD News*, 1992.

Mallory, M., & Bradford, H. (1989). An invisible workplace hazard gets harder to ignore. *Business Week*, January 30, pp. 92-93.

Morris, A. (1984). Program compliance key to preventing low back injuries. *Occupational Health & Safety*, **51**, 44-47.

Nathan, P.A., Keniston, R.C., Myers, L.D., & Meadows, K.D. (1992). Obesity as a risk factor for slowing of sensory conduction of the median nerve in industry. *Journal of Occupational Medicine*, **4**, 379-382.

National Center for Health Statistics (1979). *1979 national ambulatory medical care survey* (Series 3, No. 24, Tables 6 and 9).

National Resource Center on Worksite Health Promotion (1991). *Healthy people 2000 at work: Strategies for employers* (Appendix A). Washington, DC: Author.

Shepard, R.J. (1986). *Economic benefits of enhanced fitness*. Champaign, IL: Human Kinetics.

Shephard, R.J. (1992). A critical analysis of work-site fitness programs and their postulated economic benefits. *Medicine and Science in Sports and Exercise*, **24**, 354-370.

Shore, G., Prasad, P., & Zroback, M. (1989). Metrofit: A cost-effective fitness program. *Fitness in Business*, **3**, 147-153.

Silverstein, B.A., & Armstrong, T.J. (1988). Can in-plant exercise control musculoskeletal symptoms? *Journal of Occupational Medicine*, **30**, 922-927.

Sirles, A.T., Brown, K., & Hilyer, J.C. (1991). Effects of back school education and exercise in back injured municipal workers. *Journal of the American Association of Occupational Health Nursing*, **39**, 7-12.

Smith, B. (1990). Work-place stretching programs reduce costly accidents, injuries. *Occupational Health & Safety*, **58**, 24-25.

Stanforth, D., & Plattor, K.J. (1993). Implementing and evaluating a work station stretching program. *American Journal of Health Promotion*, **7**, 73.

Thompson, D.A. (1990). Effect of exercise breaks on musculoskeletal strain among data-entry operators: A case study. In S. Saulter et al., *Promoting Health and Productivity in the Computerized Office* (pp. 118-127). London: Taylor and Francis.

Tsai, S.P., Bernacki, E.J., & Baun, W.B. (1988). Injury prevalence and associated costs among participants of an employee fitness program. *Preventive Medicine*, **17**, 475-482.

Tsai, S.P., Gilstrap, E.L., Cowles, S.R., Waddell, L.C., & Ross, C.E. (1992). Personal and job characteristics of musculoskeletal injuries in an industrial population. *Journal of Occupational Medicine*, **34**, 606-611.

U.S. Department of Health and Human Services, Public Health Service (1990). *Healthy people 2000: National health promotion and disease prevention objectives* (DHHS Publication No. [PHS] 91-50213). Washington, DC: Author.

U.S. Department of Labor, Bureau of Labor Statistics (1992). *Newsletter*, November 18, pp. 1, 11. Washington, DC: Author.

Versloot, J.M., Rozeman, A., van Son, A.M., & van Akkerveeken, P.F. (1992). The cost-effectiveness of a back school program in industry. *Spine*, **17**, 22-27.

Walter, S.D., Sutton, J.R., & McIntosh, J.M. (1985). The etiology of sports injuries: A review of methodologies. *Sports Medicine*, **2**, 47-58.

Whitmer, W. (1992, March). The city of Birmingham's wellness partnership contains medical costs. *Business & Health*, pp. 60-66.

Chapter 8

The Impact of Worksite Health Promotion Programs on Absenteeism

William B. Baun

Absenteeism rates in the workplace have risen over 30% during the past 25 years despite the considerable improvements in both the quality of health care and socioeconomic conditions. Estimates indicated employee absence rates range from 2% to 20% across all industries, costing U.S. employers $30 billion annually (Higgins, 1988). Absenteeism has consistently plagued business and industrial organizations and has an immense literature that reflects the concern for finding both the causes of and potential solutions to this problem.

Many companies have introduced health promotion programs as a major initiative in their efforts to reduce employees' health risk and associated costs. The studies of Ogden (1987) and Golaszewski, Lynch, and Clearie (1989) provide strong evidence that employees with poor health practices have high health care costs. However, with 75% of all reported absence attributed to illness or injury, it is not surprising that there is also a strong relationship between health practices and absenteeism (Yen, Edington, & Witting, 1992). Health promotion programming may be an effective method of reducing the health risk of employees and thus decreasing the cost of poor health practices. This chapter will address the use of health promotion as one component in strategies designed to lower employee absenteeism.

In the first section of this chapter, I will discuss issues related to operational definitions in the field and provide a review of the major factors found to influence absenteeism. The second section provides a description of current absence measures. These first two sections will provide the reader with an understanding of the basic literature on absence and lay the framework for a critical review and discussion of the effect of worksite health promotion programming on absenteeism.

The third section, divided into three parts, will provide an overview of the health promotion literature and absenteeism. The first segment will provide a brief description of the studies published specific to health promotion and fitness. The second will follow with an in-depth review and discussion of the five major health promotion interventions (smoking cessation, stress reduction, hypertension control, health risk assessments, and fitness programming) found in the literature related to this topic. Finally, the last section will briefly discuss future research needs and priorities.

Absenteeism: Its Definitions and Major Factors

Employees are absent from work for reasons ranging from severe sickness or injury to the whimsical decision of just not to go to work. Ng (1988), reviewing "sickness absence" for the World Health Organization and the statistical dilemma faced by absenteeism evaluators, found sickness absence was defined as "absence from work because of sickness" or "an accident that has been accepted by the employer." In addition, sickness absence was often falsely used as an excuse for being absent. The first definition requires a verification of the problem by a medical certification, depending on the local personnel rules or insurance regulations. The second definition refers to overall absence and might include casual leave. This definition dilemma plagues the absenteeism literature and creates many challenges for those examining potential causes and solutions.

Definition Validity and Reliability

In 1980 the Educational Research Service (Porwoll, 1980) provided a summary of the employee absenteeism literature and found in a review of over 400 research articles that not one study addressed the validity of absence measures. The author also found that very few studies had computed or reported the reliability of these measures. As this research indicated, a recurring problem is the inconsistency in simply calculating absenteeism levels—what is counted as job absence in one company may differ considerably from what is counted as absence in another.

The National Center for Health Statistics (Porwoll, 1980) defined absence as "work-loss days on which a person did not work at his job or business for at least half of his normal workday because of specific illness or injury." This definition does not include work-loss due to pregnancy, family illness and injury, or health examinations. In many companies a complex set of rules is used to determine what will be a day of absence. These rules range from absence due to jury duty, disciplinary time off, death of family members, and excused absence for personal reasons that have been prearranged. Part-day absence is also defined in many different ways, often not even counted as an absence, adding more confusion to the definition dilemma.

Major Factors Influencing Employee Absence

The literature is replete with studies providing variables shown to influence an employee's absence record. Behavioral scientists have identified three major variables that are diverse but highly interrelated: personal, attitudinal, and organizational factors. Personal characteristics relating to employee absence found to be the most significant are gender, age, and occupational status (Parmeggiani, 1983; Goodman & Atkin, 1984). Although females at every age are less likely to die prematurely than males, they consistently have higher absence rates. To a greater extent these rates have been attributed to their social task as care givers in the family. Experts on aging have shown that young employees have higher absence rates, but as they approach middle age these rates decrease. After the age of 50 the second rise of absenteeism rates can be observed due to increases in debilitating diseases. Occupational status has also been shown to exert a clear influence on employee absence. For example, physical working differences and requirements between the unskilled and white collar worker has a strong influence on the ability of an employee to work with minor injuries. A sprained ankle will not allow an unskilled worker to work on a road crew, but the office worker could perform unhindered at his or her desk. It is important to recognize that many of these personal factors influence each other and the relationship they share with employee absence.

The second major factor that influences employee absence is attitudinal characteristics. Porwoll's (1980) review of employee absenteeism covering a period of 30 years suggests that dissatisfaction with work is a major determinant of employee absenteeism. This same research also indicates that the related variables of group cohesion or satisfaction is positively related to lower absenteeism and that discontentment with the direct supervisor may be a causal factor. However, several articles have shown that supervisor dissatisfaction appears to have little influence on absence rates (Hackman & Lawler, 1971; Waters & Roach, 1973; Newman, 1974; Chadwick-Jones et al., 1973).

The third major factor influencing employee absence is organizational. The most important element in this group is the size of the work unit. As the size of a work unit increases, absenteeism decreases and vice-versa (Bridges, Edwin, & Hallinan, 1978). Furthermore, the literature has shown that the type of organization also affects absenteeism rates. For example, industrial organizations have the highest absenteeism rates as compared with other employee groups (Woodall, Higgins, Dunn, & Nicholson, 1987).

All three of these factors (personal, attitudinal, organizational) must be considered when evaluating the potential causes and solutions of employee absence. Because of their strong influence on absenteeism, each should be considered when designing or reviewing pertinent studies.

Current Measures of Employee Absence

Although the three major factors discussed above provide many design challenges for researchers, the most serious problem facing the absenteeism literature concerns its measurement. In an early study by Gaudet and Frederick (1963), for example, 41 different absence measures were identified. However, the Bureau of Labor Statistics (BLS) has three standard measures used by many organizations and which are presented as guidelines here (Monthly Labor Review, 1975, 1977).

The first is incidence rate and provides a measure of the number of absences per 100 employees during a given time period (Monthly Labor Review, 1977).

$$\text{Incidence rate} = \text{number of workers absent/total employees} \times 100.$$

For example, the incidence rate of an organization that employs 250 employees and has 15 employees absent in one week would be 6%. For every 100 employees in this organization, 6 were absent during the week.

$$\text{Incidence rate} = 15/250 \times 100 = 6\%.$$

The second standard measure provided by BLS (Monthly Labor Review, 1975) is absence rate, and it provides the percent of time lost due to absence.

$$\text{Absence rate} = \text{number of hours absent/}\\ \text{number of hours usually worked} \times 100.$$

For example, if 250 employees worked 40 hours per week and each of the 15 employees absent were off work for 3 days or 24 hours, the organization's absence rate would be 3.6%. This result indicates that 3.6% of the hours usually worked were lost due to employee absence.

$$\text{Absence rate} = (15 \times 24 \text{ hours})/(250 \times 40 \text{ hours}) = 360/10{,}000 = 3.6\%.$$

The last rate measure suggested by BLS (Monthly Labor Review, 1977) is the severity rate, which provides a measure of the average time that an absent employee loses during a given period. This variable could be measured in absolute hours lost or as a percentage of usual hours worked.

$$\text{Severity rate} = \text{average number of hours lost by absent employee} / \text{average number of hours usually worked} \times 100.$$

If three employees were absent 8 hours, the severity rate would be 20% of the scheduled time, or 24 hours lost.

$$\text{Severity rate} = (3 \times 8/3 \times 40) \times 100 = 24/120 \times 100 = 20\%.$$

There are many other measurements that appear in the literature that have been reported reliable. However, many others have extreme variability and lack reliability. With so many different measures it is difficult for researchers to make valid comparisons from one study to the next. Table 8.1 lists some of the common measures used.

Worksite Health Promotion Programs and Absenteeism

Over 20 different journals have published studies concerning work absence and health promotion/fitness programming. Many of these studies have longitudinal designs where data have been collected over several years or prospectively from past time sheets or payroll records. Despite problems in quality, the ease in which absence data can be collected is probably the major reason for the abundance of literature on absenteeism. The majority of these studies have been completed on large- to middle-sized industrial and white-collar populations. Other populations studied

Table 8.1
Common Absence Measures Found in the Literature

Measure	Definition
Absence frequency	Total number of times absent
Absence severity	Total number of 1-day absences
Attitudinal absence	Frequency of 1-day absences
Medical absence	Frequency of absences of 3 days or more
Worst day absences	Difference between worst and best days absent
Blue Monday absence	Number of employees absent on a Monday less those absent on a Friday

Note. From "Absenteeism: A Review of the Literature" by P.M. Muchinsky, 1977, *Journal of Vocational Rehabilitation*, **10**, pp. 316-340.

include bus drivers, policemen, postal workers, teachers, and federal highway workers. The major health promotion programming and intervention variables studied include smoking cessation, stress management, hypertension control, health risk assessment programming, and fitness. Each of these areas will be discussed separately. Although they are not exhaustive of the studies in the absence literature, these discussions provide a good cross-section of the studies available in these areas.

Smoking Cessation and Employee Absenteeism

Smokers have consistently been found to use more health benefits (Penner & Penner, 1990; Van Peenen, Blanchard, Wolkonsky, & Gill, 1986) and sustain more occupational injuries (Cascio, 1982) than nonsmokers. Smokers also tend to have higher use of drugs (Whitehead, Smart, & Laforest, 1972) and excessive alcohol (Allen & Mazzuchi, 1985). These factors contribute to the consistently higher rates of employee absence found among smokers (Van Tuinen & Land, 1986; U.S. Dept. of Commerce, 1980; Wilson, 1973). For example, Jones, Bly, and Richardson (1990) indicated that over a 3-year period smokers had 15 hours greater annual absence than nonsmokers. The relationship between smoking and excessive absenteeism has remained consistent regardless of gender, age, or marital status (Van Tuinen & Land, 1986).

The effectiveness of smoking cessation programs has been strongly supported in the literature (Fielding, 1990; Hallet, 1986; Fisher, Glasgow, & Terborg, 1990), but does a successful program lead to lower absenteeism rates? Only a handful of smoking cessation studies have reported on absenteeism rates. Jackson, Chenoweth, Glover, Holbert, and White (1989) reported significant decreases in absenteeism in ex-smokers following a cessation program. This study was unique because it used a time-series control group and reviewed absenteeism data over a 6-year period. However, the study also examined the relationship between the number of cigarettes smoked per day and absenteeism and found that no relationship existed.

A large industrial population study reviewed the absence levels of different employee groups (managers, professionals, craft-workers, and machine operators) over a 5-year period following successful completion of smoking cessation programs (Olsen et al., 1991). Participants in the programs were found 2.3 times more likely to have quit smoking after 5 years than nonparticipants. However, success in smoking cessation had little effect on employee absenteeism. The only group that showed lower absenteeism levels were the managers who had participated in the program. The authors suggested that it may be exceedingly difficult to demonstrate the effectiveness of worksite smoking cessation programs on relatively short-term economic variables such as employee absence.

Stress and Employee Absence

Consistent evidence supports the contention that employee absenteeism increases with stress and anxiety. Douglas (1976) found that much of the reported "physical illness" used as excuses for absenteeism was really being driven by personal and environmental stress. Sylwester (1979) used a social adjustment scale and showed that those most often absent also had experienced the most change in their lives. Ramanathan (1992) found that employee stress was positively related to absenteeism and negatively related to intention to stay with the company.

A study by Hoverstad and Kjolstad (1991) used focus groups to examine employee opinions concerning the relationship between working conditions and absenteeism. When the 10 most important conditions at work were prioritized for their effect on absence, "feeling of well-being at work" was the most important factor. Several groups felt that working conditions influenced the level of absenteeism more than actual illness and that working conditions affected long- and short-term absence in the same way.

Can worksite programs designed to decrease stress and anxiety reduce absenteeism? Many worksite health promotion programs utilize Employee Assistance Programs (EAPs) to provide counseling and psychological screening for employees and dependents. EAPs can be the core provider for stress-management training. Bruhnsen (1989) reviewed the absence records for cases being seen by an EAP department and found that client absenteeism was reduced by an average of 2.87 days per year. Sixty percent of the client base reduced its reported absence by 55% and only 6% showed no change. The only job category that did not show significant drops in absenteeism was the service/maintenance group. A similar finding was shown by Ramanathan (1992) when the mean absenteeism rate was reported significantly reduced 4 months after an initial EAP contact.

Hypertension Control and Employee Absence

Hypertension has been shown to be the most common diagnosis made by physicians during office visits (Lawrence & McLemore, 1981). It has also been suggested that hypertensives have significantly higher rates of absence than individuals with normal blood pressure (Foote & Erfurt, 1983). However, Sexton and Schumann (1985) found that demographic characteristics related to absenteeism (gender, race, and age) were better determinants than hypertension. They found that only 2% of the variance in predicting work absence was accounted for by hypertension.

There has been concern in the literature that being labeled "hypertensive" can significantly increase an employee's absenteeism. In an early investigation Haynes, Sackett, Taylor, Gibson, and Johnson (1978) showed a 211% increase in illness absence in hypertensives following blood pressure screening. This raises the possibility that hypertensive labeling and not necessarily increased illness might be the real reason for increased absenteeism. However, other researchers (Rudd et al., 1987) showed that the issue of absenteeism and hypertensive labeling might be an experimental design flaw. Their 22% increase of absenteeism seen in the post-screening year was reduced significantly after using matched controls and accounting for temporal trends.

Health Risk Assessment Programs and Employee Absence

Health risk assessment (HRA) has been defined as activities designed to measure employee health status or health risk. A national survey in the early 1980s (Fielding, 1981) estimated that 29.5% of firms with 50 or more employees provided HRAs. These instruments are used to raise employee awareness of health risk and to help plan, justify, and evaluate health promotion interventions.

A constant challenge for worksite health promotion/fitness programming, in order to remain viable components of human resource departments, is to show cost-effectiveness and efficiency. HRA analysis provides an inexpensive yet cost-effective method for identification of high-risk employees. But once these employees have been identified, can they be targeted for specific behavioral change programs that will decrease their illness and absence cost?

Bertera (1991) in a large cross-sectional study showed that employees with any of six major behavioral risks (smoking, overweight, excess alcohol use, elevated cholesterol, high blood pressure, or inadequate seat belt use) had significantly higher absenteeism. The high-risk group ranged from 10% to 32% higher than the no-risk group.

In another study, Bertera (1990) implemented a health promotion program in 41 sites representing 29,315 blue-collar employees. After 2 years of programming, Bertera analyzed absenteeism specifically related to disability. Blue-collar employees at intervention sites decreased disability absence by 14%, whereas a reduction of only 5.8% was measured in the control sites. These results suggested that health promotion programming initiated with HRA and utilized within the framework of a comprehensive model can be very effective at reducing absenteeism. However, Warner (1992) questioned these findings and suggested that the absence differences observed at the end of the 2 years of programming were the exact differences observed at baseline. Bertera countered by arguing that the pretest to posttest changes were real changes of a reduction of .7 fewer disability days at program sites and .3 fewer days at nonprogram sites. These types of controversies can be found throughout the literature and underline the need for replication of results.

Fitness Programming and Employee Absence

The early economic evaluations by Cox, Shephard, and Corey (1981), Bowne (1981), and Song, Shephard, and Cox (1982) provided some of the initial findings supporting the economic justification for fitness program implementation. Many worksite programs were initiated in the 1980s, based on management expectation that medical and absenteeism cost would be reduced as soon as employees began participating. Is there a difference in the absence rates of participants and nonparticipants in fitness programs?

Baun, Bernacki, and Tsai (1986) found a significant absence difference between female participants in fitness programs (average annual hours = 47) and nonparticipants (average annual hours = 67) but no difference in males. Similar differences were found by Hoffman and Hobson (1984), whereas initial examination by Song, Shephard, and Cox (1982) showed no differences between high adherents, low adherents, and nonparticipants or dropouts.

Tucker, Aldana, and Friedman (1990) used body fat and a step-test to assess fitness levels of a large adult population. Absenteeism data were self-reported, which had been shown to be highly related to the actual reported values (Jones, Bly, & Richardson, 1990). Employees classified in the poor fitness group had more than 2.5 times the rate of absence of the high fitness group. When post hoc analyses were conducted to reduce the potential gender bias, the researchers found that the fitness/absenteeism association was stronger among females.

Using time on a treadmill as an estimate of maximal oxygen uptake, Steinhardt, Greenhow, and Stewart (1991) looked at the relationship of cardiovascular fitness to absence in a large group of police officers. Self-reported activity level was also recorded. When absenteeism was covaried for age, absenteeism was significantly related to activity level. The researchers also found that cardiovascular fitness was negatively related to absenteeism.

Boyce, Jones, and Hiatt (1991) also looked at the relationship of fitness and absenteeism in police officers. As body fat decreased by 1% in younger officers (<35 years), absenteeism increased by 0.1 day per year. In officers older than 35 years, it was found that for each one mL/kg · min increase of maximum oxygen consumption (determined by the bicycle ergometer), absenteeism decreased by 0.17 day per year. This study suggests, at least in police officers, that fitness is a poor predictor of absenteeism. Cox, Shephard, and Corey (1987) also showed a weak relationship between absenteeism and changes in maximal oxygen uptake. Taken in total, these studies suggest that fitness is weakly related to employee absence and raises doubt about the effectiveness of participation in a fitness program on absence levels.

On the other hand, Lynch, Golaszewski, Clearie, Snow, and Vickery (1990) compared participants and nonparticipants before and after establishment of a fitness center. Participants had 1.2 fewer days of absence per year. As expected, fitness center participants who had the most absence before establishment of the program showed the greatest change from frequent participation. They also reviewed absence records for a 4-year period after initiation of a communication-based program that had fitness as one of its components and found an average savings of 0.6 day per employee per year.

Absenteeism rates were also evaluated in a large metropolitan school district 1 year after program initiation (Blair, Smith, Collingwood, Reynolds, Prentice, & Sterling, 1986). With age, gender, race, and the past year's absence records as covariates, employees who participated in the program had an average of 1.25 fewer days absent during the study year when compared to nonparticipants. Wood, Olmstead, and Craig (1989) reviewed absenteeism data following 3 years of program participation and found significant differences between participants and nonparticipants (average difference of 1.86 days per year), but showed no difference within the different levels of fitness participation.

From a different perspective, fitness programs have become an important component in rehabilitation from cardiovascular incidents and repetitive movement injuries. Naas (1992) claims that the Coors cardiac rehabilitation program puts people back to work after only 2.1 months versus an average of 7.2 to 7.5 months for individuals not involved in this type of programming. Reducing the months an employee is absent from work results in a large dollar savings for the cost of replacement workers.

"Back schools" are highly accepted interventions for the reduction of worksite low back problems. But does the reduction of back pain, education on proper back working techniques, and back fitness reduce employee absence? Versloot et al. (1992) found that after initiation of a back school program with bus drivers, absenteeism levels were reduced by at least 5 days per year per employee during the 6 years the program was provided. Both these studies (Naas, 1992; Versloot et al., 1992) suggest that successful rehabilitation programs can lead to lower employee absence rates.

Conclusion

Absenteeism is a major problem for worksites, and many health promotion programs are implemented as an intervention to curb high absenteeism rates. This chapter has reviewed the health promotion literature with respect to absenteeism and, although many of the studies show positive results, the need for more and longer studies is apparent.

There is no question that smokers have higher absence rates, but the effect of smoking cessation programs on worker absence is unclear. Only a few smoking cessation studies have analyzed pre- and postparticipation absence rates, and the findings of these studies vary. Stress consistently has been shown to affect absence levels, and stress-reduction interventions (group or one-on-one counseling) have had powerful effects on reducing absence levels. Hypertension is the most common diagnosis made by a physician, and the question remains whether diagnosis alone can lead to greater absenteeism. It is unclear whether hypertensives have higher absence rates or if being labeled "hypertensive" negatively influences work attendance. Because of the few health risk assessment studies reporting absenteeism rates, it is also unclear whether HRA interventions truly affect worker absence. There are more fitness/absence studies in the literature than on any other health promotion component. These studies not only show that fitness participants are absent less often than nonparticipants, they also support the concept that the better fit are the less absent. Does fitness participation influence an individual's absence levels? The rehabilitation studies (cardiac and healthy back) suggest that fitness participation significantly affects the number of days an employee will be absent. But the relationship of fitness and absence is still unclear in the literature.

Although not definitive, the literature suggests that health habits have a significant effect on employee absenteeism. However, the question remains: Can worksites implement programs to improve health habits and thus reduce absence levels? Some of the data presented support this hypothesis, but the results are mixed. There is still a great need to replicate these studies in different work environments and company cultures.

Future Research Needs and Priorities

As long as health promotion/fitness programs claim to increase the well-being of employees, the relationship between these efforts and employee absence will remain an important indicator of program effectiveness. Unfortunately, the data on absenteeism are plagued by many measurement problems. Furthermore, the differences between voluntary and involuntary absenteeism are difficult to differentiate because of the individual biases involved.

With exceptions, absenteeism is significantly influenced by personal, attitudinal, and organizational factors, each varying within each company or worksite. This makes comparisons between data sets or between studies almost impossible. Absenteeism research needs more stable measures.

A further challenge to completing absenteeism studies is the clustering of the data. The data are generally skewed with many individuals having little or no absence and a few individuals accumulating most of the total absenteeism. This makes correlation studies difficult and can reduce the power of significance tests. Based on the above, future research should consider

- a standardized definition of absence (perhaps what is advocated by the Bureau of Labor Standards),

- a longer period of study (3 to 5 years) with larger population samples, and
- statistical analysis adjusting for confounding factors that also affect absences (e.g., gender, age, and occupation status).

References

Allen, J., & Mazzuchi, J. (1985). Alcohol and drug abuse among American military personnel: Prevalence and policy implications. *Military Medicine*, **150**, 250-255.

Baun, W., Bernacki, E., & Tsai, S. (1986). A preliminary investigation: Effect of a corporate fitness program on absenteeism and health care cost. *Journal of Occupational Medicine*, **28**, 18-22.

Bertera, R.L. (1990) The effects of workplace health promotion on absenteeism and employment costs in a large industrial population. *American Journal of Public Health*, **80**, 1101-1105.

Bertera, R.L. (1991). The effects of behavioral risks on absenteeism and health care costs in the workplace. *Journal of Occupational Medicine*, **33**, 1119-1124.

Blair, S., Smith, M., Collingwood, T., Reynolds, M., Prentice, M., & Sterling, C. (1986). Health promotion for educators: Impact of absenteeism. *Preventive Medicine*, **15**, 166-175.

Boyce, R., Jones, G., & Hiatt, A. (1991). Physical fitness capacity and absenteeism of police officers. *Journal of Occupational Medicine*, **33**, 1137-1143.

Bowne, D.W. (1981). Physical fitness programs for industry—an extravagance or a wise investment? *Transactions of the Association of Life Insurance Medical Directors of America*, **64**, 210-222.

Bridges, E., Edwin, M., & Hallinan, M. (1978). Subunit size, work system interdependence, and employee absenteeism. *Educational Administration Quarterly*, **14**, 24-42.

Bruhnsen, K. (1989). EAP evaluation and cost benefit savings: A case example. *Health Values*, **13**, 39-42.

Cascio, W.F. (1982). Costing the effects of smoking in the work place. In W.F. Cascio (Ed.), *The Financial Impact of Behavior in Organizations* (pp. 65-78). Boston: Kent.

Chadwick-Jones, J., Brown, C., & Nicholson, N. (1973). Absence from work: Its meaning, measurement, and control. *Industrial Review of Applied Psychology*, **22**, 137-155.

Cox, M., Shephard, R., & Corey, P. (1981). Influences of an employee fitness programme upon fitness, productivity and absenteeism. *Ergonomics*, **24**, 795-806.

Cox, M., Shephard, R., & Corey, P. (1987). Physical activity and alienation in the work place. *Journal of Sports Medicine and Physical Fitness*, **27**, 429-436.

Douglas, S.A. (1976). Social-psychological correlates of teacher absenteeism—a multi-variate study. (Doctoral dissertation, Ohio State University, Columbus). *Dissertation Abstracts International*, **37**, 11A.

Fielding, J.E. (1981). Frequency of health risk assessment activities at U.S. worksites. *American Journal of Preventive Medicine*, **5**, 73-81.

Fielding, J.E. (1990). Worksite health promotion survey: Smoking control activities. *Preventive Medicine*, **19**, 402-413.

Fisher, K.J., Glasgow, R.R., & Terborg, J.R. (1990). Worksite smoking cessation: A meta-analysis of long-term quit rates from controlled studies. *Journal of Occupational Medicine*, **32**, 429-439.

Foote, A., & Erfurt, J. (1983). *Hypertension control in the work setting* (NIH Publication 83-2012). Hyattsville, MD: National Center for Health Statistics.

Gaudet, L., & Frederick, J. (1963). *Solving the problems of employee absence* (AMA Research Study 57). New York: American Management Association.

Golaszewski, T., Lynch, W., & Clearie, A. (1989). The relationship between retrospective health insurance claims and a health risk appraisal-generated measure of health status. *Journal of Occupational Medicine*, **31**, 262-264.

Golaszewski, T., Snow, D., Lynch, W., Yen, L., & Solomita, D. (1992). A benefit-to-cost analysis of a work-site health promotion program. *Journal of Occupational Medicine*, **34**, 1164-1172.

Goodman, P., & Atkin, R. (1984). *New approaches to understanding, measuring, and managing employee absence*. San Francisco: Jossey-Bass.

Hackman, J.R., & Lawler, E.E. (1971). Employee reactions to job characteristics. *Journal of Applied Psychology*, **55**, 259-286.

Hallet, R. (1986). Smoking intervention in the workplace: Review and recommendations. *Preventive Medicine*, **15**, 213-231.

Haynes, R.B., Sackett, K.L., & Taylor, D.W. (1981). Increased absenteeism from work after detection and labeling of hypertensive patients. *New England Journal of Medicine*, **299**, 741-744.

Higgins, C.W. (1988). The economics of health promotion. *Health Values*, **12**, 39-45.

Hoffman, J., & Hobson, C. (1984). Physical fitness and employee effectiveness. *Personnel Administration*, **4**, 101-126.

Hoverstad, T., & Kjolstad, S. (1991). Use of focus groups to study absenteeism due to illness. *Journal of Occupational Medicine*, **33**, 1046-1050.

Jones, R.C., Bly, J.L., & Richardson, J.E. (1990). A study of a work site health promotion program and absenteeism. *Journal of Occupational Medicine*, **32**, 95-99.

Jackson. S.E., Chenoweth, D., Glover, E.D., Holbert, D., & White, D. (1989, December). Study indicates smoking cessation improves workplace absenteeism rate. *Occupational Health & Safety*, pp. 13-18.

Lawrence, L., & McLemore, T. (1981). *1981 summary: National ambulatory medical care survey* (PHS No. 83-1250). Hyattsville, MD: National Center for Health Statistics.

Lynch, W.D., Golaszewski, T.J., Clearie, A.F., Snow, D., & Vickery, D.M. (1990). Impact of a facility-based corporate fitness program on the number of absences from work due to illness. *Journal of Occupational Medicine*, **32**, 9-12.

Monthly Labor Review (1975, August). *Unscheduled absence from work—an update*, **98**, pp. 36-39.

Monthly Labor Review (1977, October). *Absence from work—measuring the hours lost*, **100**, pp. 16-23.

Naas, R. (1992). Health promotion programs yield long-term savings. *Business and Health*, **10**(13), 41-47.

Newman, J.E. (1974). Predicting absenteeism and turnover: A field comparison of Fishbein's model and traditional job attitude measures. *Journal of Applied Psychology*, **59**, 610-615.

Ng, T.K. (1988). Descriptive occupational morbidity statistics. In *Proceedings from the 1988 International Conferences of Labour Statisticians* (pp. 200-204).

Ogden, D.F. (1987). *Health risks and behavior: The impact on medical costs* (preliminary study by Milliman & Robertson, Inc. and Control Data). Unpublished internal report.

Olsen, G.W., Lacy, S.E., Sprafka, J.M., Arceneaux, T.G., Potts, T.A., Kravat, B.A., Gondex, M.R., & Bond, G.G. (1991). A 5-year evaluation of a smoking cessation incentive program for chemical employees. *Preventive Medicine*, **20**, 774-784.

Parmeggiani, L. (Ed.) (1983). *Encyclopaedia of occupational health and safety* (3rd ed.). Geneva: International Labour Office.

Penner, M., & Penner, S. (1990). Excess insured health care costs from tobacco-using employees in a large group plan. *Journal of Occupational Medicine*, **32**, 521-523.

Porwoll, P.J. (1980). *Employee absenteeism: A summary of research*. Arlington, VA: Educational Research Services.

Ramanathan, C.S. (1992). EAP's response to personal stress and productivity: Implications for occupational social work. *Social Work*, **37**, 234-239.

Rudd, P., Price, M., Graham, L., Beilstein, B., Tarbell, S., Bacchetti, P., & Fortmann, S. (1987). Consequences of worksite hypertension screening: Changes in absenteeism. *Hypertension*, **10**, 425-436.

Sexton, M., & Schumann, B. (1985). Sex, race, age, and hypertension as determinants of employee absenteeism. *American Journal of Epidemiology*, **122**, 302-310.

Song, T.K., Shephard, R.J., & Cox, M. (1982). Absenteeism, employee turnover and sustained exercise participation. *Journal of Sports Medicine*, **22**, 392-399.

Steinhardt, M., Greenhow, L., & Stewart, J. (1991). The relationship of physical activity and cardiovascular fitness to absenteeism and medical care claims among law enforcement officers. *American Journal of Health Promotion*, **5**, 455-460.

Sylwester, R. (1979). Educator absences and stress. *OSSC Quarterly Report*, **19**, 18-21.

Taylor, D.W., Haynes, R.B., Sackett, D.L., & Gibson, E.S. Longterm follow-up of absenteeism among working men following detection and treatment of their hypertension. *Clinical Investigative Medicine*, **4**, 173-177.

Tucker, L.A., Aldana, S.G., & Friedman, G.M. (1990). Cardiovascular fitness and absenteeism in 8,301 employed adults. *American Journal of Health Promotion*, **5**, 140-145.

U.S. Department of Commerce, Bureau of the Census (1980). *Statistical abstract of the United States* (101st ed.). Washington, D.C.: Author.

Van Peenen, P., Blanchard, A., Wolkonsky, P., & Gill, T. (1986). Health insurance claims of petrochemical company employees. *Journal of Occupational Medicine*, **28**, 237-240.

Van Tuinen, M., & Land, G. (1986). Smoking and excess sick leave in a department of health. *Journal of Occupational Medicine*, **28**, 33-35.

Versloot, J.M., Rozeman, A.M., van Son, A.M., & van Akkerveeken, P.F. (1992). The cost-effectiveness of a back school program in industry. *Spine*, **17**, 22-27.

Warner, K.E. (1992). Effects of workplace health promotion not demonstrated. *American Journal of Public Health*, **82**, 126.

Waters, L.K., & Roach, D. (1973). Job attitude as predictors of termination and absenteeism: Consistency over time and across organizational units. *Journal of Applied Psychology*, **57**, 341-342.

Whitehead, P.C., Smart, R.G., & Laforest, L. (1972). Multiple drug use among marijuana smokers in eastern Canada. *International Journal of Addiction*, **7**, 179-190.

Wilson, R.W. (1973). Cigarette smoking, disability days, and respiratory conditions. *Journal of Occupational Medicine*, **15**, 236-240.

Wood, E.A., Olmstead, G.W., & Craig, J.L. (1989). An evaluation of lifestyle risk factors and absenteeism after two years in a worksite health promotion program. *American Journal of Health Promotion*, **4**, 128-133.

Woodall, G.E., Higgins, C.W., Dunn, J.D., & Nicholson, T. (1987). Characteristics of the frequent visitor to the industrial medical department and implications for health promotion. *Journal of Occupational Medicine*, **29**, 660-664.

Yen, L.T., Edington, D.W., & Witting, P. (1992). Prediction of prospective medical claims and absenteeism costs for 1284 hourly workers from a manufacturing company. *Journal of Occupational Medicine*, **34**, 428-435.

Chapter 9

Worksite Health Promotion and Productivity

Roy J. Shephard

Productivity is here considered as the average output of goods and/or services for each hour a worker is at work. Sickness, absenteeism, and employee turnover each has an important impact on the efficiency of a company's operations, but these items are considered in other chapters.

If an employee is working maximally, one might anticipate a hyperbolic relationship between the quality and the quantity of the worker's output (see Figure 9.1). The objective of the successful manager is thus to optimize the quality/quantity relationship for a minimum input of personnel, materials, and/or equipment. In the context of the present volume, the question that arises is how far can a worksite fitness and health promotion program assist the manager in optimizing productivity? In practice, it may be difficult for either the manager or the exercise scientist to monitor the quality of output, and productivity is often measured much more simply. For example, in a physical task such as fighting a forest fire, the observer may determine the length of firebreak cut per hour (Danielson & Danielson, 1982), and in an office job such as an airline reservation or insurance clerk, one may judge the number of telephone calls answered or the policies processed per hour relative to the standard time allowed for this particular task (Cox, Shephard, & Corey, 1981).

Potential gains of productivity figure fairly prominently in the stated rationale of companies that have already introduced fitness and health promotional programs; moreover, such gains seem of even greater interest to those who are contemplating the introduction of a health and fitness program (see Table 9.1 on page 149). Nevertheless, claims of an association between the introduction of worksite wellness programs and increased productivity often lack clear empirical support (Gibbons, 1989; Haskell, 1988; Pencak, 1991), and any linkage between enhanced fitness and greater

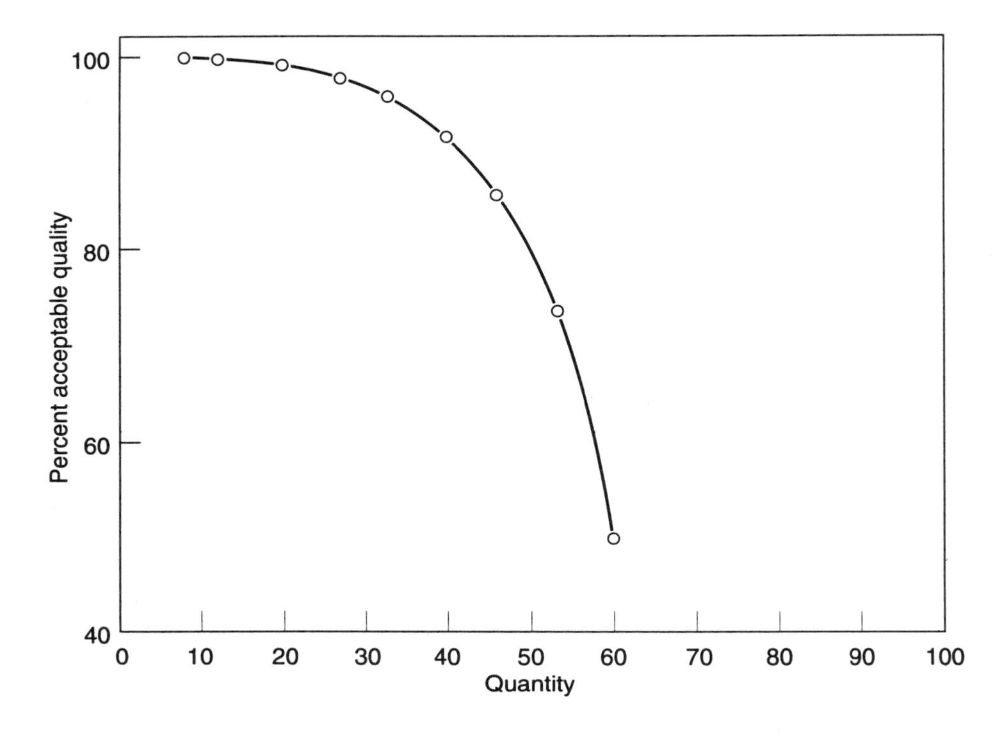

Figure 9.1 Relationship between quantity and quality of production at the worksite. In many operations, there is also a minimum quality accepted by management and/or consumers.
Note. From *Men at Work. Applications of Ergonomics to Performance and Design* by R.J. Shephard, 1974, Springfield, IL: CC Thomas.

productivity seems quite tenuous. Potential intervening variables include enhanced health, greater job satisfaction, higher levels of personal energy, and less fatigue (Howard & Michalachki, 1979).

This chapter will examine the theoretical reasons why fitness and health promotion programs might modify worker productivity, and it will explore the likely impact of such programs on the various determinants of occupational performance. Empirical evidence on gains of productivity will then be critically reviewed, and a final brief section will discuss the policy implications for management and investigators.

Mechanisms Increasing Worker Productivity

At first inspection, the practical benefits from a fitness and health promotion program seem obvious in a task that demands heavy physical work.

Table 9.1
Reasons for Current Operation or Interest in Development of Corporate Health Promotion Programs

Stated reason	Existing program	Interest in developing program
Employee health	82%	68%
Employee morale	59	52
Health costs	57	67
Turnover/absenteeism	51	57
Productivity	50	64
Employee demand	33	20
Innovative trend	32	11
Corporate image	20	18

Note. Adapted from data in *Worksite Health Promotion in Colorado* by M.F. Davis, K. Rosenberg, D.C. Iverson, T.M. Vernon, and J. Bauer, 1984, U.S. Public Health Reports, 99, p. 540.

If employees are initially working at a physiological limit imposed by their aerobic power, their muscle strength, or their heat tolerance, then a training-induced gain in the mechanical efficiency of working, or an increase in maximal oxygen intake, strength, or heat tolerance might be expected to yield a matching gain of production. Early investigators seem to have accepted the truth of this hypothesis without experimental proof. For example, Pravosudov (1978) assumed that because aerobic power was greater in Soviet "worker-athletes" than in their sedentary peers, the active individuals necessarily had a higher worksite productivity.

However, on closer analysis, the factors limiting occupational productivity are complex (Shephard, 1986a,b). Worksite constraints may include not only the abilities, motivation, and experience of the individual employee, but also the cooperation of key colleagues, the efficiency of management, any restrictions on the rate of working imposed by the union or trade association, the availability of raw materials and equipment, the overall quality of the working environment, capital investment in the automation of repetitive tasks, and consumer demand for the product. Many of these items are not susceptible to the physical efforts of the individual worker, or even of a small team of workers. The impact of a program on the quality or quantity of production will thus depend on whether immediate human resources form the rate- or the quality-limiting step in the productive process.

We will now examine this proposition with respect to both physical and mental tasks.

Physical Tasks

If a person's physical performance characteristics are marginal from the viewpoint of task demands, then the quantity of production will diminish for self-paced work, and in machine-paced situations excessive fatigue may lead to poor quality output or a loss of production from industrial disputes, accidents, injuries, absenteeism, and employee turnover. How commonly does a physiological limitation arise in terms of aerobic power, muscle strength, or thermoregulation?

Aerobic Power. Estimates based on the performance of self-paced heavy work suggest that an individual can sustain a load 40% of his or her maximal oxygen intake over an 8-hour shift (Hughes & Goldman, 1970). However, the assessment of specific tasks is complicated by the fact that the 40% ceiling is situational. Values can vary from 35% to 50% of maximal oxygen intake, depending on the nature and circumstances of the required task (Åstrand, 1967; Bonjer, 1966). Acceptable percentages diminish with adoption of an awkward posture, the use of small muscle groups, intermittent peaks of more intensive physical demand, or adverse environmental conditions (Shephard, 1974). Much also depends on the individual's motivation: Much higher percentages of peak effort are accepted in an athletic competition or in an emergency than during normal routine industrial work.

Even if we adopt 40% of aerobic power as a useful and realistic criterion, it is surprisingly difficult to assess how many physically demanding jobs approach this ceiling of effort. Theoretically, the energy cost of specific worksite tasks can be evaluated quite accurately by indirect calorimetry, using a device such as a Kofranyi-Michaelis respirometer or an Oxy-log recorder. But, in practice, employees commonly accelerate their work pace when the equipment is worn, so that the data obtained are not representative of normal operations. Further, on a normal workday, the hypothesized sequence of occupational tasks is usually performed for much less than a full 8-hour shift, although both management and unions may be reluctant to admit the extent of "lost time."

A priori, a physiological limitation does not seem very likely in most jobs (Howard & Michalachki, 1979). Through many years of experience and/or union negotiations, tools (e.g., shovels and wheelbarrows) and tasks have been designed so that the occupational demand is held below the aerobic ceiling for an average member of the labor force. Ergonomists commonly operate with the concept of a relaxation allowance (Shephard, 1974); the standard pace is set to demand 80% of the nominal potential of an average worker (that is, 32% of maximal oxygen intake).

The standard pace was commonly agreed on when the majority of those engaged in heavy work were young men. Let us assume an average worker, 40 years of age, continued employment through the age of 65. This worker would experience a 10% loss of aerobic power per decade between 40 and 65 years of age (Shephard, 1987a), with a 30 to 40% difference of aerobic power between men and women (Shephard, Vandewalle, Bouhlel, & Monod, 1988). We may then anticipate that the average 60-year-old male and many female workers (both young and old) will encounter difficulty in meeting the aerobic demands of heavy jobs. In practice, complaints are not frequent even among older workers, for several reasons, including the following:

- The original 80% standard was reached by negotiation between the employer and the employees and is thus less than the true, physiologically imposed ceiling.
- Female (and to a lesser extent male) employees who choose heavy occupations have physical working capacities above the population average.
- In many tasks, the functional determinant is not the individual's absolute working capacity but rather his or her relative working capacity (where women are at a smaller disadvantage).
- The experience gained through performance of a given task over many years increases an employee's mechanical efficiency, so that the energy cost of this particular task is less for older, more experienced members of the labor force.
- Participation in physically demanding work may help to sustain aerobic fitness and/or muscle strength as a person ages.
- Job categorizations are crude and, because of promotion or seniority rules, many of those classed as "heavy workers" have been able to select less demanding work as they become older.
- Individuals who find that because of aging their working capacity or health is no longer equal to their task commonly seek transfer to alternative employment or early retirement.

Muscle Strength. Dynamic muscle strength seems a more important limiting factor than aerobic power in many types of heavy work. Wyndham (1966) quickly discovered that in the difficult physical conditions of some South African mines, body mass (probably serving as a surrogate for lean tissue mass) provided the best method of selecting productive laborers. Nottrodt and Celentano (1984) noted that even among young recruits to the Canadian armed services, a substantial minority of men and most women were unable to meet the required task demands of front-line service (lifting to shoulder height a load of 18 kg repeatedly and of 35 kg occasionally). Given the anticipated decrease in muscle strength of at least 25% over a working career (Shephard, 1987a), it seems likely that muscle strength and/or endurance might become an important factor limiting the performance of many demanding physical tasks, and that productivity in such jobs might be enhanced by measures designed to increase muscle strength.

Thermoregulation. Heat stress may be imposed by either the working environment or the need to wear protective clothing. Productivity suffers (Barrett, 1991) mainly because of personal discomfort (Berglund, 1988). Likely consequences are a low level of worker satisfaction, poor quality output, industrial disputes, and accidents. As core temperature rises, the peak cardiac output is reduced because fluid is lost in sweating and an increasing fraction of the total blood flow is diverted from the working muscles to the skin; the fraction of maximal oxygen intake that can be used without fatigue thus decreases. In some tasks such as deep mining, physical demands may even push core temperatures to levels where heat collapse and heat stroke are possible (Wyndham & Strydom, 1972).

Training interacts with frequent heat exposure, generally reducing these stresses, but there remain many heavy physical tasks where thermoregulation is a major concern, with a potential impact on productivity.

Conclusions Concerning Physical Tasks. Many industrial physiologists have focused on aerobic power as the factor limiting output in heavy physical work. In occupations that require much walking with heavy loads (for instance, foot soldiers and mail carriers), this may be the main constraint (Hughes & Goldman 1970; Shephard, 1982). However, in many occupations, an inadequate muscle strength or problems of thermal regulation are more likely to impose constraints on worksite productivity.

Mental Tasks

In mentally demanding tasks, the quantity and/or quality of production may fall because a job is boring. Alternatively, work may impose heavy emotional stress for reasons such as excessive task difficulty, a lack of control or perceived control over operations, unclear job requirements, personality conflicts, or inadequate feedback of performance.

In the case of boring work, a fitness program might serve as an arousing stimulus, bringing the subject back to an optimal point on the inverted-U curve relating arousal and performance (see Figure 9.2). In contrast, if the worker is overstressed, some types of recreation might offer relaxation, again bringing the individual back to the peak of the inverted-U relationship (Martens, 1974; Norris, Carroll, & Cochrane, 1990; Pauly et al., 1982), with an improvement of performance on the job.

General Benefits

We may finally note some more general mechanisms whereby fitness and health promotion programs might have a favorable influence on productivity.

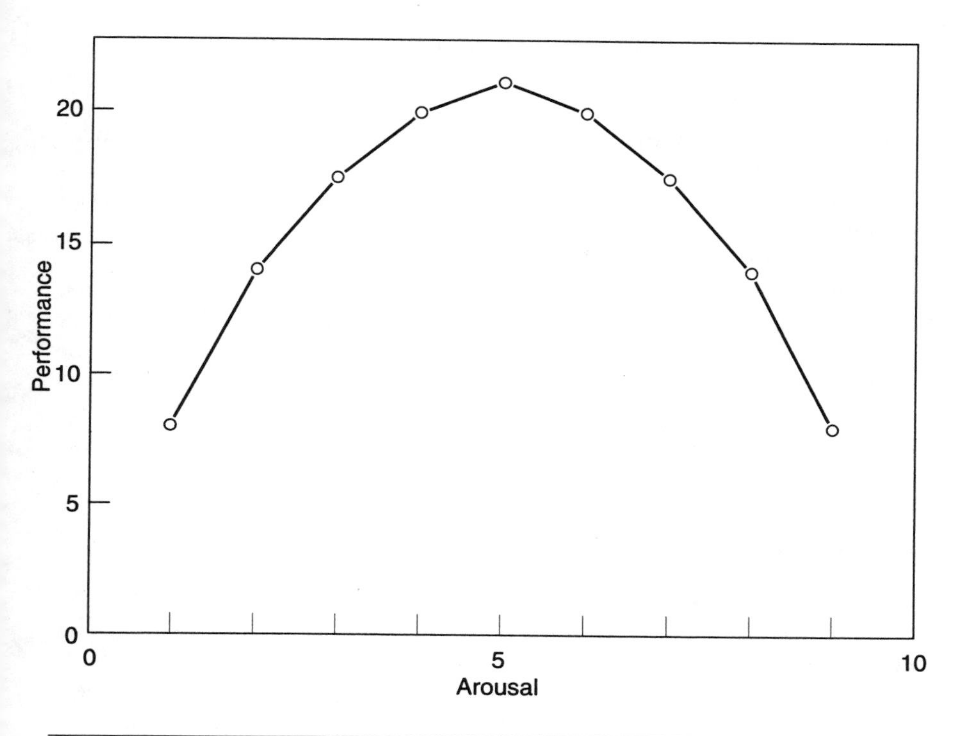

Figure 9.2 The inverted-U relationship between arousal and productivity. Note that in a repetitive, boring job, production is constrained because arousal is less than optimal, whereas in a stressful job, productivity is impaired by excessive arousal.

Workers might perceive such initiatives as an expression of management interest in their personal welfare (Spilman et al., 1986) with a resulting increase in job satisfaction (Howard & Michalachki, 1979). In such situations, management might be successful in recruiting premium employees (Bernacki & Baun, 1984). Moreover, current employees might be prepared to work closer to their physical or mental limits, to cooperate better with each other, to take greater care of equipment and materials, and to engage in fewer industrial disputes.

Fitness programs might also enhance the worker's self-image, and because of a sense of greater self-efficacy (Bandura, 1977), an employee might perform the required tasks more effectively.

Finally, wellness programs might influence aspects of personal lifestyle other than physical fitness (Henritze, Brammell, & McGloin, 1992; Wood, Olmstead, & Craig, 1989). For example, if smoking withdrawal is encouraged, then less time is lost through the immediate mechanics of smoking (Kristein, 1982) or because employees leave the worksite repeatedly to visit an area where smoking is permitted.

Impact of Worksite Programs on Health and Function

Having identified some theoretical mechanisms whereby worksite fitness and health promotion programs might influence productivity, we must next examine the likely effectiveness of such programs in an industrial setting. The impact depends immediately on the content of the program. One immediate difficulty in interpreting the current literature is that often little or no information is given about the nature of the fitness, sports, recreational, and wellness programs that have been instituted. Sometimes programs have included substantial elements of health promotion or health education. In such instances, employees have often had the opportunity to select one or more options from a modular program, obscuring details on participation rates, cost, and the nature of the treatment actually received.

Constraints on Program Impact

Impact at the worksite is often greatly attenuated relative to the changes observed when the same program is applied in a laboratory setting. The attenuation of benefits may conveniently be illustrated by reference to the likely gains of aerobic power. If a group of employees is persuaded to attend a regular laboratory program of endurance training, one may anticipate an average increase in maximal oxygen intake of perhaps 20% (Shephard, 1977). However, the response at the worksite is much less dramatic. Typically, only 20% of employees are initially recruited to a fitness program, and perhaps 10% continue to comply with program requirements beyond the first few months (Leatt et al., 1988; Shephard, 1986a,b, 1987b). The response observed in the 10% of continuing participants is attenuated relative to laboratory findings because the training stimulus is smaller (training sessions are shorter, attendance is less consistent, and in some instances the program is less demanding or less effective). Moreover, some of the most conscientious exercisers have already participated in fitness programs elsewhere (Leatt et al., 1988), and others already have a fairly high initial fitness level because of the physical demands of their employment. Thus, the 20% gain of maximal oxygen intake observed in laboratory volunteers may be reduced to an overall, company-wide benefit of 1% or less if a similar program is introduced at the worksite.

Aerobic Power. Empirical data support the above predictions concerning aerobic power. The impact of a worksite fitness program is much less than would be anticipated in the laboratory. Slee and Peepre (1974) and Cox et al. (1981) both found a 13% increase of predicted aerobic power in response to office fitness programs, but gains were restricted to frequent program participants, only about 10% of the total work force. Thus, the corporate impact was only a 1.3% gain of maximal oxygen intake. The populations that they examined were desk workers, so aerobic power was unlikely to have limited job performance in any direct fashion. But even if the work had been physically demanding, and the aerobic power of the oldest third of the labor force had initially been inadequate to meet job demands, the observed increment of oxygen transport could not have boosted productivity by more than about 0.4% (1.3 × 0.33%).

Two years after the inception of Johnson and Johnson's Live for Life campaign, Blair et al. (1986a) claimed to have achieved a larger (10.5%) gain in maximal oxygen intake, averaged across all employees at participating worksites. However, it was necessary to estimate gains from submaximal cycle ergometer test data, and control subjects who had received only a health screening also showed a 4.7% increase of fitness score at the final evaluation (as a result either of habituation to the test procedures or of a change in lifestyle induced by screening). The net program effect was thus a 5.8% rather than a 10.5% gain of aerobic power. Moreover, the program effect relative to control subjects was even smaller (0.9%) in the over 45-year-old employees, where one might have anticipated problems of productivity due to a lack of aerobic power.

Muscle Strength. If older employees were willing to spend 30 minutes once or twice per week in muscle-strengthening exercises, there is little doubt that their ability to lift heavy loads repeatedly could be improved substantially (Fiatarone et al., 1990). Some worksite facilities have purchased expensive commercial devices for resisted muscle training. Such programs supposedly improve simple strength test scores and reduce the incidence of back injuries (Barnard & Anthony, 1980; Superko, Bernauer, & Voss, 1983). Slee and Peepre (1974) found a significant (4.6%) increase of grip strength but no significant increase of scores for explosive power (standing broad jump) after office workers had participated in a thrice-weekly worksite aerobics program for 4 months. In contrast, Cox et al. (1981) found a small (3.9%) decrease in grip strength (in men only) but a substantial (10%) increase in lower back flexibility in both sexes in response to a similar program.

Danielson and Danielson (1982) carried out a controlled study on forest fire fighters. The experimental group received a daily running, calisthenics, and games program. Relative to control crews, they demonstrated gains in maximal oxygen intake (12.8%), Sargent jump test (5.8%), and sit-up scores (8.4%), but the grip strength of the active group actually deteriorated (4.5% loss).

Given that the problem of inadequate strength is particularly marked among female employees, it is discouraging that until recently many women who join worksite fitness classes have had little interest in weight-lifting types of exercise programs (Peepre, 1978).

Thermoregulation. There is some interaction between aerobic training and heat acclimatization (Gisolfi, 1973), and those who improve their maximal oxygen intake through a worksite fitness program might thus be expected to show a somewhat greater tolerance of extreme environmental conditions. However, for maximal benefit, subjects must be exposed to a combination of heat and regular exercise. Wyndham (1974) has used this technique to prepare new recruits for deep gold mining in South Africa.

The impact of any enhanced fitness on the employee's tolerance of more moderate heat is less clearly established. Thermal comfort is in large measure determined by skin-wettedness (Berglund, 1988), and because fit individuals sweat more readily, training may not improve the level of comfort if employees must work under conditions where it is difficult for sweat to evaporate (Aoyagi, McLellan, & Shephard, 1994).

General Effects of Programs

It might be anticipated that in addition to specific physiological changes, the introduction of extensive fitness and health promotion facilities and programs would have more general effects on overall work satisfaction, feelings of self-efficacy and mood-state, the level of arousal and/or relaxation of employees, and their adoption of healthy lifestyles. The validity of these assumptions has been tested and will be discussed here.

Overall Work Satisfaction. It might seem logical that companies providing extensive fitness and health promotion facilities would be able to recruit premium employees and have a labor force with a high level of work satisfaction (Rudman, 1987). Baun et al. (1986) argued that companies with a "fitness image" selectively recruited those individuals interested in fitness and that such people tended to be high achievers. Their main evidence for this hypothesis was a cross-sectional comparison of performance and job turnover rates between participants and nonparticipants in a company fitness program (Bernacki & Baun, 1984). Retrospective analysis suggested that high performance antedated participation in the fitness program. Even if the advantage of selective recruitment were substantiated by more direct experiments, gains would depend on the immediate labor market and the working conditions offered by rival employers.

Formal measures of work satisfaction have shown little difference between fitness program participants and control groups (Cox, Shephard, & Corey, 1987; Cox & Montgomery, 1991; Rudman, 1987). Indeed, in some instances workers have invested sufficient time in the mechanics of organizing and/or participating in company fitness and sports activities for this to have a negative overall impact on productivity (Golaszewski et al., 1992; Shephard, 1974). The test instruments used (for instance, the job satisfaction index of Smith, Kendall, & Hulin, 1969) may not have had sufficient sensitivity to detect small program-related gains in work satisfaction. Furthermore, the companies evaluated to date (typically long-established, white-collar, nonunion operations) sometimes have had a high initial level of worker satisfaction (King, Murray, & Atkinson, 1981), leaving little scope for improvement. A final possibility is that workers may have had difficulty in establishing a connection between the fitness/health promotion program and immediate causes of dissatisfaction such as poor salary and promotion prospects (Cox et al., 1987). Nevertheless, in one questionnaire survey, 69% of Canadian workers believed that the introduction of a company-based employee fitness program would boost their morale (Wanzel, 1974). Likewise, Rudman (1987) found such a belief was prevalent among the employees of Campbell's Soups in the United States.

Even if further and better-designed experiments were to show an increase of work satisfaction in response to an employee fitness or wellness program, the long-term impact of such initiatives on motivation and productivity would be debatable. Those who have used monetary payments to increase productivity have long recognized that the response is maximized by rewards that are variable in amount and administered on a random schedule (Shephard, 1974; Skinner, 1953). An exercise or health promotion session, consistently available in a predictable fashion every noon hour, thus compares unfavorably with the alternative stimulus of an occasional and variable monetary bonus.

Increased Self-Efficacy and Elevation of Mood-State. The impact of fitness programs on the participants' feelings of self-efficacy is well documented (Bandura, 1977; Godin & Shephard, 1990; Long, 1984). Likewise, many exercisers suggest that the reason they participate in an exercise program is because they "feel better," and meta-analyses support the concept that aerobic exercise programs can be helpful in correcting anxiety and depression (Castell & Blumenthal, 1985; Martinsen, 1990; Petruzzello et al., 1991). However, the extent of any response depends on the individual's initial level of fitness and the initial lack of self-efficacy or disturbance of mood-state (Eickhoff, Thorland, & Ansgorge, 1983; Jasnoski & Holmes, 1981).

Given the highly subjective nature of occupational fatigue, an increase of self-efficacy and/or an elevation of mood-state are likely to allow the completion of demanding tasks without complaints of tiredness. However, reports of the impact of worksite fitness programs on self-efficacy have been limited to subjective (and generally uncontrolled) accounts of more positive attitudes toward work, with decreased feelings of strain and tension (Barker, 1988; Durbeck et al., 1972; Halfon et al., 1990; Heinzelmann, 1975; Itoh et al., 1990; Landgreen, 1988; Okada & Iseki, 1990), and (in one popular and uncontrolled report) an increase of police officer commendations (Mealey, 1979). Moreover, in at least a few instances there has been either no increase of job satisfaction (Cox et al., 1987; Cox & Montgomery, 1991) or a company-wide improvement of attitudes affecting both participants and nonparticipants in the health and fitness program (Rudman, 1987).

Increased Arousal/Increased Relaxation. The typical worksite fitness program, whether a brief "fitness break" or a formal exercise class, seems likely to increase arousal (Bahrke & Morgan, 1981). Early observations of the response (see LaPorte, 1970, and Shephard, 1991, for references) detailed favorable subjective reactions, increased attention, faster visual responses and flicker fusion frequency, improvements in visual acuity, and a decrease in errors during electrical assembly, textile inspection, manual telegraph operation, and data entry.

In contrast, it is a matter of common experience that some types of physical activity, particularly recreation in the countryside, can contribute to relaxation (Bahrke & Morgan, 1981). A meta-analysis of cross-sectional studies suggests that aerobically fit individuals have lesser physiological responses to "stress," with faster recovery rates than their sedentary peers (Crews & Landers, 1987). Keller and Serganian (1984) further suggested that a well-designed exercise program was more effective than either meditation or music appreciation classes as a means of speeding electrodermal recovery from acute stressful situations. Durbeck et al. (1972) and Heinzelmann (1975) made controlled comparisons of exercisers and sedentary subjects. Self-reports indicated positive changes of work performance in some 60% of experimental subjects but only 3% of controls. Gains (related to program participation and aerobic fitness) included a more positive self-image, with associated increases in stamina (both physical and mental), greater energy, improved powers of concentration and decision-making, and a greater ability to deal with stress—these benefits developing over the course of an 18-month aerobic conditioning program. Self-reports claimed similar benefits when NASA employees were recruited to an onsite fitness facility (Heinzelmann, 1975). However, there remains a need to obtain further objective data in controlled trials with average worksite fitness classes.

Healthy Lifestyle. The association between participation in physical activity and adoption of a healthy lifestyle is not necessarily very strong (Shephard, 1989). Much depends on the type of physical activity undertaken. The influence on overall health behavior is greater for endurance-type programs than for social games and sports such as tennis. The context of program delivery is also likely to be significant, with a potential for greater impact if associated health promotion and wellness modules are open to exercise class participants.

From the viewpoint of productivity loss, smoking and substance abuse are of particular interest. The maximum likely benefit from an endurance-type program is a doubling of successful cigarette withdrawal (Morgan,

Gildiner, & Wright, 1976), with small additional benefits from alcohol withdrawal (Wilbur, 1983). Unfortunately, heavy consumers of cigarettes and alcohol are not commonly attracted to or retained by fitness programs (Massie & Shephard, 1971). If we assume an overall exercise participation rate of 20%, that 20% of participants are initially smokers, and that 50% of those initially addicted are cured by the exercise class, then the overall program impact is that 2% of the labor force will stop smoking. Self-reports from the Toronto Life Assurance study (3% smoking withdrawal in first year; Shephard et al., 1982), the Staywell Program (6% decrease of smokers relative to controls at year 3; Anderson & Jose, 1987), and Johnson and Johnson (3.5% smoking withdrawal in one year, 12% in 2 years; Wilbur, 1983) support this estimate of effect magnitude.

In a computer simulation of findings from the Adolph Coors Wellness Program, Terborg (1988) estimated productivity losses at 1.8% for smokers and 30% for alcoholics. Thus, if exercise were to augment productivity by reducing the number of smokers in a company, it would at most induce a gain of 0.22% (12% $\times$ 0.018).

Evidence of Enhanced Productivity

This chapter does not consider sickness, absenteeism, or employee turnover, although these are important tangible components of lost productivity (Gettman, 1986). Not only is there a loss of production if a person is absent from work, but a new recruit or a temporary replacement may operate much less efficiently than a fully trained employee.

Measurement Problems

An objective assessment of productivity would examine the output of goods and services per worker in relation to the quality of the product. However, observers commonly have been content to report subjective, intangible items: self-reports of self-confidence, morale and productivity, supervisor ratings, commendations, or merit pay awards. Even when such reports have been favorable, they could have been biased upward by an improvement of mood-state or by "halo" effects, without any change of actual productivity. Likewise, the assessments of supervisors are susceptible to their opinions about the program, management directives regarding the program, and any perceived side effects such as injuries. Thus, Blair et al. (1980) found a small negative impact of self-selected leisure on supervisor ratings in white female insurance clerks, but this effect disappeared when the data were covaried to allow for the greater sick-leave that had been taken by the active group.

Any perceived benefit from an employee fitness and health promotion program should ideally be evaluated by a double-blind controlled experiment. However, there are many practical obstacles to the conduct of such an assessment (see Kirkcaldy & Shephard, 1990, for detailed review). The most important source of difficulty is that a fitness and health promotion program cannot be introduced into a company without both the employees and those responsible for their "objective" assessment being aware of the intervention. This leaves the experiment very vulnerable to halo and Hawthorne effects (Pennock, 1930). The treated group tends to develop positive expectations and also enjoys regular contact with a concerned health advisor, whereas the controls may develop a negative feeling because they have been "neglected" (Garfield, 1981). One possible experimental tactic (exploited in the laboratory but not yet in industry) would be to introduce alternative programs of varying intensity or hypothesized effectiveness at different worksites and to test for associated gradients in productivity.

The few studies of productivity in which control groups have been arranged have generally received a minor intervention, such as a health examination, a routine fitness assessment, or lifestyle counseling. In the study of Cox et al. (1981), there was a 4.3% gain of productivity at the control site, presumably mainly a Hawthorne-type response to the added attention (participation in a major experiment and opportunity for a free fitness assessment). At the experimental company, productivity was increased by 7% in the year following introduction of an employee fitness and lifestyle program. It could be argued that the 2.7% difference between the two companies reflected a true program effect, but it is unsatisfactory that this gain was not observed uniformly across the company and that the added response seen at the experimental company was smaller than the nonspecific Hawthorne effect at the control site. It thus remains possible that a larger and more persistent intervention at the experimental company resulted in a larger Hawthorne effect at this location than at the control company. Certainly, individual supervisor ratings did not show any strong relationship between program participation and increased productivity (Shephard, Cox, & Corey, 1981).

Empirical Data

Empirical data on productivity are summarized in Table 9.2 on pages 162 through 164. Unfortunately, few of the studies have adopted strong experimental designs. In no instance have subjects been allocated randomly between experimental and control groups. A few authors have sought matching control worksites. But in many instances comparisons have merely been drawn with subjects who elected not to participate in the fitness and health promotion program, or no attempt has been made to find any control data. In one "analysis" (Golaszewski et al., 1992), a 4% gain in productivity was assumed merely from a superficial reading of the literature! Often program participation rates are poorly documented and the content of the experimental treatment is unclear. Finally, measures of productivity have frequently been subjective and qualitative rather than objective and quantitative.

Heavy Physical Tasks. In some instances, benefit has apparently accrued because the fitness program has enhanced aerobic power or muscle strength. In other instances, a lessening of fatigue has been observed (LaPorte, 1970) or imputed (Rohmert, 1973) because a fitness break allowed relaxation from a primary task that demanded intensive static muscle efforts. Participants in exercise classes have frequently reported enhanced physical stamina and a lessening of fatigue. Given the demonstrated program-related increases of aerobic power, it is possible such comments have an objective, biological basis, although part of the change could also arise from a positive attitude of the subjects toward the experimental intervention, or as an indirect consequence of an elevation of mood-state.

Danielson and Danielson (1982) carried out one of the few controlled studies of a heavy physical task. The productivity of forest fire fighters was assessed on the objective basis of the rate of cutting of a firebreak (meters cut per crew member per hour). On average, the productivity of experimental (aerobically trained) crews was 62% higher than that of control crews. For any given level of physiological function, productivity was positively correlated with the worker's age. After covariance adjustment for age and experience, performance remained significantly correlated with the individual's maximal oxygen intake per kg of body mass. However, when the data were subdivided in terms of weather conditions, the correlation between productivity and aerobic power persisted only under hot summer conditions, when the fire hazard was rated as high. In this type of situation, the trained group was able to maintain its work output relative to easier conditions, and thus developed a 101% advantage of cutting speed relative to control subjects. After allowance for the effects of age and experience, muscle strength had no positive influence on productivity, and indeed Burpee scores were negatively correlated with performance (probably because the individuals concerned were heavy, muscular, and had a low relative aerobic power).

Light Physical Work. Uncontrolled but objective studies have suggested that fitness programs induce decreases of error scores in such jobs as manual telegraph operation and the inspection of textiles, with an increase in the speed of unpaced manual tasks. A controlled study in the Israeli pharmaceutical industry (Halfon et al., 1990; Rosenfeld et al., 1990) compared responses to 7 months of worksite physical or social activity; physical activity was associated with decreased self-perceptions of work load and fatigue and enhanced work satisfaction. Nevertheless, all of these

Table 9.2
Empirical Data on the Relationships Between Productivity and the Introduction of Fitness and Health Promotional Programs

Study	Topic	Subject/study site	Methodology
Anderson (1991)	Productivity related to entry physical ability test scores	Grocery warehouse	Cross-sectional study
Barker (1988)	Increased satisfaction with supervision, job satisfaction, organizational commitment	Johnson & Johnson	Matched control sites
Bernacki & Baun (1984)	Association of fitness and performance ratings	Tenneco	Cross-sectional data antedated program introduction
Blair et al. (1980)	Negative impact on supervisor ratings (only in white women, not seen if covaried for absenteeism)	Liberty Life Insurance	
Blair et al. (1984)	Marginal self-reported gains, due mainly to increased well-being	Teachers	Three tests, one control school
Blair et al. (1986b)	Gains of physical self-concept, energy level, well-being, freedom from worry, life satisfaction, mood, tenseness, emotional control, *but* small decrease of job satisfaction	Dallas school board	Participants vs. nonparticipants 10-week study only
C. Cox & Montgomery (1991)	No relation between submaximal cycle ergometer score and job satisfaction	Hospital employees	Cross-sectional data
M. Cox et al. (1981)	2.7% advantage over control company	Canada Life Assurance	Large Hawthorne effect
Croce & Horvat (1992)	Aerobic and resistance-based program increased pieces of work completed	Mentally retarded	Uncontrolled, three obese men

Danielson & Danielson (1982)	Twice as long firebreak in trained workers	Ontario forest fire fighters	Benefit only seen in hot and difficult conditions
Durbeck et al. (1972)	Self-reports—work enjoyed more, less boring, able to work harder	NASA HQ	No controls, but gains correlated with participation and fitness
Golaszewski et al.	Anecdotal improved morale	Travelers Insurance	No data, no controls
Halfon et al. (1990)	Reduction of fatigue and perceived effort	Pharmaceutical workers	Social recreational control
Health & Welfare Canada	4% gain of data processing productivity	Government office	No controls
Heinzelmann (1975)	52% increase of self-reported productivity	NASA employees	Nonparticipant controls
Howard & Michalachki (1979)	No gain of productivity	Middle managers	No details of study published
Karch et al. (1988)	Supervisors estimated increased productivity; participants estimated no change	Pentagon staff	No controls, some data on nonparticipants
LaPorte (1970)	Gains of strength, hand steadiness; less fatigue of eye muscles	Postal workers	Passive break controls
Manguroff et al. (1960)	20% to 30% increase of speed	14 professional tasks	No controls
Mealey (1979)	39% increase of commendations	Police officers	No controls
Norris et al. (1990)	Self-reports of well-being and job stress reduced more by aerobic than by anaerobic training	Police officers	Controlled study
Oden (1988) Oden et al. (1989)	Statistically nonsignificant gain in job satisfaction over 24 weeks of aerobic training; no change in quality or quantity of production (but previously underemployed)	Westinghouse	Controlled study (N = 45)
Okada & Iseki (1990)	70% increase in gas sales per worker	Gas Utility company	No controls
Pauly et al. (1982)	Decrease in anxiety, improved personal, social and physical self-concept	Xerox Corp.	Correlation analysis Gains unrelated to participation

(continued)

Table 9.2 *(continued)*

Study	Topic	Subject/study site	Methodology
Réville	30% reduction of errors	Textile workers	No controls
Ricci (1992)	Self-reports of increased productivity, more energy, less stress, better self-image, more self-confidence, better concentration, greater enthusiasm for work	Electric company office	No controls
Rudman (1987)	Companywide increased job satisfaction	Campbell's Soups	Self-selected membership
Shephard et al. (1981)	Supervisory ratings improved similarly in experimental and control companies	Canada Life Assurance	Matched control company
Shinew & Crossley (1988)	900 randomly selected employees. Participants in recreation and fitness programs have higher job satisfaction, less absenteeism	General Electric Co.	Comparison with nonparticipants
Shoenhair & Stone (1988)	Increased productivity	Elec. executives	Matched controls, small study
Slee and Peepre (1974)	Self-reports of enhanced stamina, less stress and tension, better work output	Government office	Nonparticipant controls
Smolander et al. (1992)	5.5% gain in "work ability index" relative to 2.5% in controls	Metal company	Controls; abstract only
Spilman et al. (1986)	Perception of company interest in welfare. No gains of work enthusiasm or satisfaction with working conditions	AT&T	Matched control sites
Stallings et al. (1975)	No effect on teaching or research	U.S. faculty	Self-selected sport vs. sedentary faculty
Terborg (1988)	0% to 1% gain of productivity	Coors Brewery. Theoretical model	Sensitivity analysis
Zoltick et al. (1990)	5.6% gain of reported productivity	Pentagon	No controls

findings could be no more than halo or Hawthorne-type responses or indirect effects of an exercise-induced improvement in mood-state.

Mental Work. Some subjective reports have claimed very large gains in productivity as a result of participation in health and fitness programs (Barker, 1988; Durbeck et al., 1972; Heinzelmann, 1975; Karch et al., 1988; Ricci, 1992; Shoenhair & Stone, 1988; Zoltick et al., 1990). For example, in the study of Slee and Peepre (1974), there were self-reports of greater stamina (71% of female and 64% of male participants), a more positive attitude to work (53% of females and 29% of males), less stress and tension (78% of females and 36% of males), improved work performance (59% of females and 29% of males), increased self-confidence (60% of females and 36% of males), and improved self-image (65% of females and 62% of males). The gender differences in this particular trial were attributed to differences in participation rates and seniority between women and men. However, such self-reports can be biased by improved general well-being (Blair et al., 1984), favorable mood changes, and positive attitudes toward a fitness and lifestyle program. Further, in some groups that are probably well-satisfied with their initial status (for instance, middle managers and university professors), self-reports have shown no gains in ratings (Cox & Montgomery, 1991; Howard & Michalachki, 1979; Rudman, 1987; Stallings, O'Rourke, & Gross, 1975) or even a decrease in work satisfaction (Blair et al., 1986b) following program implementation. The issue may be further clouded by a selective loss of workers from fitness and health programs, as those dissatisfied with company pay policy (often the major cause of job dissatisfaction) are likely either not to participate or to drop out of fitness and health promotion programs (Shephard & Cox, 1980).

There have been some reports of gains in objectively measured productivity following the introduction of fitness programs (e.g., Health & Welfare Canada, 1976; Okada & Iseki, 1990; Zoltick et al., 1990), but these studies have been uncontrolled. Cox et al. (1981) adopted the objective tactic of comparing work times to previously established standard times for specified tasks. In the first year after establishment of the experimental program, this approach demonstrated a 7.0% gain of productivity at the experimental company versus a 4.3% gain at the control company. However, it is disturbing that although observations were made at 23 departments in the experimental company, there was a gain in productivity of 31% in 3 departments and in the remaining 20 departments the observed benefit of 3.4% was less than at the control site. The absence of any major effect relative to controls was confirmed by Shephard et al. (1981). They obtained objective supervisor ratings for six aspects of performance (cooperation with workmates and supervisor, accuracy of work, promptness of arrival, job satisfaction, and productivity). Over the first 6 months of the experimental program, scores showed a 6% improvement at both experimental and control sites. Likewise, Smolander et al. (1992) assessed a "work ability index" in a controlled study of a blue-collar operation (a metal company). The exercised subjects showed a 5.5% gain in performance after program implementation, but there was a disturbingly large 2.5% improvement in control subjects.

Implications for Policy

From the foregoing review, it is premature to conclude that a fitness and health promotion program will necessarily boost a person's productivity at the worksite, and it would certainly be rash to speculate on the magnitude of any effect. The first step should be to carry out better designed experiments to evaluate the impact of well-specified programs on the performance of both physical and mental work. This will require the development of good indices of individual and corporate productivity, particularly in companies that do not have a clearly defined physical end-product. Given that exercise and lifestyle programs cannot be administered in double-blind fashion, there will also be a need for alternative placebo interventions that have an equal impact on participant attitudes and expectations. Future studies should compare objective measures of productivity at experimental and placebo sites, preferably using a single-blind randomized design. If early gains of productivity are seen at the experimental sites, it will remain important to assess whether these are a short-term reaction to an innovative treatment or a more permanent response to program-induced gains of fitness and health. It will further be necessary to test whether effects are demonstrated most readily in companies where work satisfaction is initially low. Finally, if companies continue to view fitness and health promotion programs as a significant means of boosting productivity, it will be necessary to compare any responses demonstrated with other potential tactics for achieving the same goals. Alternative approaches might range from investment in automation through an upgrading of the overall working environment and an increase of salaries to various techniques for optimization of the worker's arousal level.

References

Anderson, C.K. (1991). Physical ability testing as a means to reduce injuries in grocery warehouses. *International Journal of Retail Distribution Management*, **19**(7), 33-35.

Anderson, D.R., & Jose, W.S. (1987). Employee lifestyle and the bottom line. Results from the Staywell Evaluation. *Fitness in Business*, **2**, 86-91.

Aoyagi, Y., McLellan, T., & Shephard, R.J. (1994). Effects of endurance training on exercise-heat tolerance in men wearing nuclear, biological and/or chemical protective clothing. *European Journal of Applied Physiology*.

Åstrand, I. (1967). Degree of strain during building work as related to individual aerobic capacity. *Ergonomics*, **10**, 293-303.

Bahrke, M.S., & Morgan, W.P. (1981). Anxiety reduction following exercise and meditation. In M.H. Sacks & M.L. Sachs (Eds.), *Psychology of Running*. Champaign, IL: Human Kinetics.

Bandura, A. (1977). Self-efficacy: Toward a unifying theory of behavioral change. *Psychological Review*, **84**, 191-215.

Barker, F. (1988). Worksite health promotion—today and tomorrow. In H. Myers (Ed.), *Decreasing Barriers: A Blueprint for Workplace Health in the Nineties* (pp. 5-16). Dallas: American Heart Association.

Barnard, R.J., & Anthony, D.F. (1980). Effect of health maintenance programs on Los Angeles City firefighters. *Journal of Occupational Medicine*, **22**, 667-669.

Barrett, M.V. (1991). Heat stress disorders. Old problems, new implications. *American Association of Occupational Health Nurses Journal*, **39**, 369-380.

Baun, W.B., Bernacki, E.J., & Tsai, S.P. (1986). A preliminary investigation: Effect of a corporate fitness program on absenteeism and health care cost. *Journal of Occupational Medicine*, **28**, 18-22.

Berglund, L.G. (1988). Thermal comfort: A review of some recent research. In I.B. Mekjavic, E.W. Banister, & J.B. Morrison (Eds.), *Environmental Ergonomics* (pp. 70-86). London: Taylor & Francis.

Bernacki, E.J., & Baun, W.B. (1984). The relationship of job performance to exercise adherence in a corporate fitness program. *Journal of Occupational Medicine*, **30**, 949-952.

Blair, S.N., Blair, A., Howe, H., Russell, P., Rosenberg, M., & Parker, G. (1980). Leisure time physical activity and job performance. *Research Quarterly*, **51**, 718-723.

Blair, S.N., Collingwood, T.R., Reynolds, R., Smith, M., Hagan, D., & Sterling, C.S. (1984). Health promotion for educators. *American Journal of Public Health*, **74**, 147-149.

Blair, S.N., Piserchia, P.V., Wilbur, C.S., & Crowder, J.H. (1986a). A public health intervention model for work-site health promotion. *Journal of the American Medical Association*, **255**, 921-926.

Blair, S.N., Smith, M., Collingwood, T.R., Reynolds, R., Prentice, M.C., & Sterling, C.L. (1986b). Health promotion for educators: Impact on absenteeism. *Preventive Medicine*, **15**, 166-175.

Bonjer, F.H. (1966). Relationships between physical working capacity and allowable caloric expenditure. In W. Rohmert (Ed.), *Muskelarbeit und Muskeltraining* (pp. 86-98). Darmstadt: Gentner Verlag.

Castell, P.J., & Blumenthal, J.A. (1985). The effects of aerobic exercise on mood. In B. Kirkcaldy (Ed.), *Individual Differences in Movement*. Lancaster & Boston: MTP.

Cox, C.L., & Montgomery, A.C. (1991). Fitness and absenteeism among hospital workers. *American Association of Occupational Health Nurses Journal*, **39**, 189-198.

Cox, M.H., Shephard, R.J., & Corey, P.N. (1981). Influence of an employee fitness programme upon fitness, productivity and absenteeism. *Ergonomics*, **24**, 795-806.

Cox, M.H., Shephard, R.J., & Corey, P.N. (1987). Physical activity and alienation in the work place. *Journal of Sports Medicine and Physical Fitness*, **27**, 429-436.

Crews, D.J., & Landers, D.M. (1987). A meta-analysis review of aerobic fitness and reactivity to psychosocial stressors. *Medicine and Science in Sports and Exercise*, **19**, 114-120.

Croce, R., & Horvat, M. (1992). Effects of reinforcement-based exercise on fitness and work productivity in adults with mental retardation. *Adapted Physical Activity Quarterly*, **9**, 148-178.

Danielson, R., & Danielson, K. (1982). Exercise program effects on productivity of forestry fire fighters. Toronto: Ontario Ministry of Tourism and Recreation.

Durbeck, D.C., Heinzelmann, F., Schacter, J., Haskell, W.L., Payne, G.H., Moxley, R.T., Nemiroff, M., Limoncelli, D.D., Arnoldi, L.B., & Fox, S.M. (1972). The National Aeronautics and Space Administration—U.S. Public Health Service Evaluation and Enhancement Program. *American Journal of Cardiology*, **30**, 784-790.

Eickhoff, J., Thorland, W., & Ansgorge, C. (1983). Selected physiological and psychological effects of aerobic dancing among young adult women. *Journal of Sports Medicine*, **23**, 273-280.

Fiatarone, M.A., Marks, E.C., Ryan, N.D., Meredith, C.N., Lipsitz, L.A., & Evans, W.J. (1990). High intensity strength training in nonagenarians. Effects on skeletal muscle. *Journal of the American Medical Association*, **263**, 3029-3034.

Garfield, S.L. (1981). Critical issues in the effectiveness of psychotherapy. In C.E. Walker (Ed.), *Clinical Practice of Psychology*. Oxford: Pergamon Press.

Gettman, L.R. (1986, July). Cost/benefit analysis of a corporate fitness program. *Fitness in Business*, **1**, 11-17.

Gibbons, L.W. (1989). Corporate fitness programmes and health enhancement. *Annals of the Academy of Medicine, Singapore*, **18**, 272-278.

Gisolfi, C. (1973). Work-heat tolerance derived from interval training. *Journal of Applied Physiology*, **35**, 349-354.

Godin, G., & Shephard, R.J. (1990). Use of attitude-behaviour models in exercise promotion. *Sports Medicine*, **10**, 103-121.

Golaszewski, T., Snow, D., Lynch, W., Yen, L., & Solomita, D. (1992). A benefit to cost analysis of a work-site health promotion program. *Journal of Occupational Medicine*, **34**, 1164-1172.

Halfon, S.T., Rosenfeld, O., Ruskin, H., & Tenenbaum, G. (1990). Daily physical activity program for industrial employees. In M. Kaneko (Ed.), *Fitness for the Aged, Disabled and Industrial Worker* (pp. 260-265). Champaign, IL: Human Kinetics.

Haskell, W.L. (1988). Exercise as a means of maximizing human physical performance and productivity. In R.S. Williams, & A.G. Wallace (Eds.), *Biological Effects of Physical Activity* (pp. 115-126). Champaign, IL: Human Kinetics.

Health and Welfare, Canada (1976). *Your lifestyle profile—operation lifestyle.* Unpublished Report on Employee Fitness.

Heinzelmann, F. (1975). Psychosocial implications of physical activity. In S. Keir (Ed.), *Employee Physical Fitness in Canada* (pp. 33-43). Ottawa, ON: Information Canada.

Henritze, J., Brammell, H.L., & McGloin, J. (1992). Lifecheck: A successful, low-tech, in plant, cardiovascular disease risk identification and modification program. *American Journal of Health Promotion*, **7**, 129-136.

Howard, J.H., & Michalachki, A. (1979). Fitness and employee productivity. *Canadian Journal of Applied Sport Sciences*, **4**, 191-198.

Hughes, A.L., & Goldman, R.F. (1970). Energy cost of hard work. *Journal of Applied Physiology*, **29**, 570-572.

Itoh, I., Hanawa, K., & Okuse, S. (1990). Exercise program for increasing health and physical fitness of workers. In M. Kaneko (Ed.), *Fitness for the Aged, Disabled and Industrial Worker* (pp. 245-248). Champaign, IL: Human Kinetics.

Jasnoski, M.L., & Holmes, D.S. (1981). Influence of initial aerobic fitness, aerobic training and changes in aerobic fitness on personality functioning. *Journal of Psychosomatic Research*, **25**, 553-556.

Karch, R.C., Newton, D.L., Schaeffer, M.A., Zoltick, J.M., Zajtchuk, R., & Rumbaugh, J.H. (1988). *Cost-benefit and cost-effectiveness measures of health promotion in a military/civilian staff.* Technical Report to the Department of the Army, Office of the U.S. Surgeon General.

Keller, S., & Serganian, P. (1984). Physical fitness level and autonomic reactivity to psychosocial stress. *Journal of Psychosomatic Research*, **28**, 279-287.

King, M., Murray, M., & Atkinson, T. (1981). Background personality, job characteristics and satisfaction with work in a national sample. *Canadian Journal of Behavioural Sciences*, **13**, 44-52.

Kirkcaldy, B., & Shephard, R.J. (1990). Therapeutic implications of exercise. *International Journal of Sport Psychology*, **21**, 165-184.

Kristein, M.M. (1982). The economics of health promotion at the work site. *Health Education Quarterly*, **9**(Suppl.), 27-36.

Landgreen, M. (1988). Health management: Moving from corporate headquarters to all employees; a larger corporation's perspective. In H. Myers (Ed.), *Decreasing Barriers: A Blueprint for Workplace Health in the Nineties* (pp. 43-50). Dallas: American Heart Association.

LaPorte, W. (1970). *La gymnastique de pause dans l'entreprise*. Brussels: Editions de L'administration de L'education Physique des Sports.

Leatt, P., Hattin, H., West, C., & Shephard, R.J. (1988). Seven year follow-up of employee fitness program. *Canadian Journal of Public Health*, **79**, 20-25.

Long, B.C. (1984). Aerobic conditioning and stress inoculation: A comparison of stress management interventions. *Cognitive Therapeutic Research*, **8**, 517-542.

Martens, R. (1974). Arousal and motor performance. *Exercise and Sport Science Reviews*, **2**, 155-188.

Martinsen, E.W. (1990). Benefits of exercise for the treatment of depression. *Sports Medicine*, **9**, 380-389.

Massie, J.F., & Shephard, R.J. (1971). Physiological and psychological effects of training. *Medicine and Science in Sports*, **3**, 110-117.

Mealey, M. (1979). New fitness for police and firefighters. *Physician in Sportsmedicine*, **7**(7), 96-100.

Morgan, P.W., Gildiner, M., & Wright, G.R. (1976). Smoking reduction in adults who take up exercise: A survey of running clubs for adults. *Canadian Association for Health, Physical Education and Recreation Journal*, **42**(5), 39-43.

Norris, R., Carroll, D., & Cochrane, R. (1990). The effects of aerobic and anaerobic training on fitness, blood pressure, and psychological stress and well-being. *Journal of Psychosomatic Research*, **34**, 367-375.

Nottrodt, J.W., & Celentano, E.J. (1984). Use of validity measures in the selection of physical screening tests. In D.A. Attwood & C. McCann (Eds.), *Proceedings of the 1984 International Conference on Occupational Ergonomics* (pp. 433-437). Toronto: Human Factors Society of Canada.

Oden, G.L. (1988). The effects of an employee fitness program on worker productivity, absenteeism and health care cost. Doctoral dissertation, Texas A&M University, College Station.

Oden, G., Crouse, S.F., & Reynolds, C. (1989). Worker productivity, job satisfaction, and work stress: The influence of an employee fitness program. *Fitness in Business*, **3**(6), 198-203.

Okada, K., & Iseki, T. (1990). Effects of a ten year corporate fitness program on employees' health. In M. Kaneko (Ed.), *Fitness for the Aged, Disabled and Industrial Worker* (pp. 249-253). Champaign, IL: Human Kinetics.

Pauly, J.T., Palmer, J.A., Wright, C.C., & Pfeiffer, G.J. (1982). The effect of a 14 week employee fitness program on selected physiological and psychological parameters. *Journal of Occupational Medicine*, **24**, 457-463.

Peepre, M. (1978). *Employee fitness and lifestyle report*. Ottawa, ON: Fitness and Amateur Sport Directorate.

Pencak, M. (1991). Workplace health promotion programs. An overview. *Nursing Clinics of North America*, **26**, 233-240.

Pennock, G.A. (1930). Investigation of rest periods, working conditions and other influences. *Personnel Journal*, **8**. Cited by M.E. Mundel (1970). *Motion and Time Study. Principles & Practice*. Englewood Cliffs, NJ: Prentice Hall.

Petruzzello, S.J., Landers, D.M., Hatfield, B.D., Kubitz, K.A., & Salazar, W. (1991). A meta-analysis on the anxiety-reducing effects of acute and chronic exercise. Outcomes and mechanisms. *Sports Medicine*, **11**, 143-182.

Pravosudov, V.P. (1978). Effects of physical exercise on health and economic efficiency. In F. Landry & W.A.R. Orban (Eds.), *Physical Activity and Human Well-being* (pp. 261-271). Miami, FL: Symposia Specialists.

Réville, P.H. (1970). *Sport for all. Physical activity and the prevention of disease*. Strasbourg: Council of Europe.

Ricci, J. (1992, May). Perceived benefits of a 9-month, community-based, corporate fitness program. Poster presentation at the *International Conference on Physical Activity, Fitness, and Health*, Toronto.

Rohmert, W. (1973). Problems in determining rest allowances. Part I. Use of modern methods to evaluate stress and strain in static muscular work. *Applied Ergonomics*, **4**, 91-95.

Rosenfeld, O., Tenenbaum, G., Ruskin, H., & Halfon, S-T. (1990). Behavioural modifications following a physical activity programme in the Israeli pharmaceutical industry. *Australian Journal of Sports Science*, **22**, 93-96.

Rudman, W.J. (1987). Do onsite health and fitness programs affect worker productivity? *Fitness in Business*, **2**(1), 2-8.

Shephard, R.J. (1974). *Men at work. Applications of ergonomics to performance and design*. Springfield, IL: CC Thomas.

Shephard, R.J. (1977). *Endurance fitness* (2nd ed.). Toronto: University of Toronto Press.

Shephard, R.J. (1986a). *Fitness and health in industry*. Basel: Karger Publications.

Shephard, R.J. (1986b). *Economics of enhanced fitness*. Champaign, IL: Human Kinetics.

Shephard, R.J. (1987a). *Physical activity and aging*. London: Croom Helm.

Shephard, R.J. (1987b). Exercise adherence in corporate settings: Personal traits and program barriers. In R. Dishman (Ed.), *Exercise Adherence—Its Impact on Public Health* (pp. 305-319). Champaign, IL: Human Kinetics.

Shephard, R.J. (1988). Sport, leisure and well-being: An ergonomics perspective. *Ergonomics*, **31**, 1501-1517.

Shephard, R.J. (1989). Exercise and lifestyle change. *British Journal of Sports Medicine*, **23**, 11-22.
Shephard, R.J. (1991). A short history of occupational fitness and health promotion. *Preventive Medicine*, **20**, 436-445.
Shephard, R.J., & Cox, M.H. (1980). Some characteristics of participants in an industrial fitness programme. *Canadian Journal of Applied Sport Sciences*, **5**, 69-76.
Shephard, R.J., Cox, M.H., & Corey, P.N. (1981). Fitness program participation; its effects on worker performance. *Journal of Occupational Medicine*, 359-363.
Shephard, R.J., Corey, P.N., & Cox, M.H. (1982). Health hazard appraisal—the influence of an employee fitness programme. *Canadian Journal of Public Health*, **73**, 183-187.
Shephard, R.J., Vandewalle, H., Bouhlel, E., & Monod, H. (1988). Sex differences of physical work capacity in normoxia and hypoxia. *Ergonomics*, **31**, 1177-1192.
Shinew, K.J., & Crossley, J.C. (1988). A comparison of employee recreation and fitness program benefits. *Employee Benefits Journal*, **13**(4), 20.
Shoenhair, C.L. (1987). *The influence of long-term exercise on absenteeism, health-care costs and productivity*. Unpublished master's thesis, Arizona State University, Tempe.
Shoenhair, C.L., & Stone, W.J. (1988, May). The influence of long-term exercise on absenteeism, health care costs and productivity. Poster presentation at the *International Conference on Exercise, Fitness and Health*, Toronto.
Skinner, B.F. (1953). *Science and human behavior*. New York: MacMillan.
Slee, D., & Peepre, M. (1974). *Report on health and welfare pilot employee fitness program*. Ottawa, ON: Fitness and Amateur Sport Branch.
Smith, P.C., Kendall, L.M., & Hulin, C.L. (1969). *The measurement of satisfaction in work and retirement*. Chicago: Rand, McNally.
Smolander, J., Louhevaraa, V., Ilmarinen, J., & Korhonen, O. (1992, May). Feasibility and effects of a work-site exercise promotion program in a metal company. Poster presentation at the *International Conference on Physical Activity, Fitness, and Health*, Toronto.
Spilman, M.A., Goetz, A., Schult, J., Bellingham, R., & Johnson, D. (1986). Effects of a corporate health promotion program. *Journal of Occupational Medicine*, **28**, 285-289.
Stallings, W.M., O'Rourke, T.W., & Gross, D. (1975). Professorial correlates of physical exercise. *Journal of Sports Medicine and Physical Fitness*, **15**, 333-336.
Superko, H.R., Bernauer, E., & Voss, J. (1983). Effects of a mandatory job performance test and voluntary remediation program on law enforcement personnel. *Medicine and Science in Sports and Exercise*, **15**, 149-150.

Terborg, J.R. (1988). *Cost/benefit analysis of the Adolph Coors wellness program* (Unpublished Report). Eugene, OR: University of Oregon College of Business Administration.

Wanzel, R.S. (1974). *Determination of attitudes of employees and management of Canadian corporations toward company-sponsored physical activity facilities and programs*. Unpublished doctoral dissertation, University of Alberta, Edmonton.

Wilbur, C.S. (1983). The Johnson & Johnson program. *Preventive Medicine*, **12**, 672-681.

Wood, E.A., Olmstead, G.W., & Craig, J.L. (1989). An evaluation of lifestyle risk factors and absenteeism after two years in a work-site health promotion program. *American Journal of Health Promotion*, **4**, 128-133.

Wyndham, C.H. (1966). An examination of the methods of physical classification of African labourers for manual work. *South African Medical Journal*, **40**, 275-278.

Wyndham, C.H. (1974). 1973 Yant Memorial Lecture. Research in the human sciences in the gold mining industry. *American Industrial Hygiene Journal*, **35**, 113-136.

Wyndham, C.H., & Strydom, N.B. (1972). Körperliche Arbeit bei hoher Temperatur. In W. Hollmann (Ed.), *Zentrale Themen der Sport Medizin* (pp. 131-149). Berlin: Springer Verlag.

Zoltick, J.M., Karch, R.C., Newton, D.L., Schaeffer, M.A., Zajtchuk, R., & Rumbaugh, J.H. (1990). Health promotion in a military/corporate setting. *Medicine and Science in Sports and Exercise*, **22**, S44.

HAROLD BRIDGES LIBRARY
S. MARTIN'S COLLEGE
LANCASTER

Chapter 10

Health Benefits Design

Thomas J. Golaszewski

This chapter will review the emerging area of health benefits design, with an emphasis on the use of benefits as incentives and its relationship to the mission of health promotion. Unlike many more established areas in the field of worksite health promotion, there is a dearth of available information on this subject. With this limited history, a brief review of definitions and conceptual parameters is in order.

Benefits are defined as virtually any type of compensation or services provided by the employer in any form other than direct wages (Employee Benefit Research Institute, 1987). Some are required by law such as employer contributions to Social Security or Medicare, but most are discretionary. Because benefits contribute to the economic well-being of individuals, they function as part of the compensation package; hence they serve to recruit, reward, and maintain employees. Subsequently, benefits vary from organization to organization depending on the competitiveness of the marketplace and the financial status of the company.

In contrast, health promotion is defined as "any combination of educational, organizational, economic, and environmental support for behavior conducive to health" (Green & Johnson, 1983). Its purpose is to influence health behavior and, by this definition, health promotion is clearly aggressive. All ethical means of influencing health behavior are permissible, including mandates by law or, in the case of organizations, mandates by policy.

Although the relationship hasn't been fully exploited, health promotion and traditional benefits (health and life insurance, sick leave, etc.) have the potential to be inextricably linked. Benefits exercise an enormous and growing influence on employee behavior. Usually, this influence is channeled in the context of job selection, "Do I prefer to work for company A or company B?" But this same influence can be used to shape employee behavior in a variety of other contexts. For example, because of the growing value attached to health insurance, modifications in the benefits package can be used to meet other needs of the organization, including those related to improving employee health and well-being.

With consideration for the above, this chapter will first review background information examining factors contributing to the current health care cost crisis in industry. Second, the topic of incentives will be explored as they relate to worksite health promotion in general. Finally, a discussion of companies that use benefit-related incentives specifically will be provided, including an examination of program features, anecdotal observations, and, where available, evaluation results.

With respect to definitions, Chenoweth (1988) argues for the distinction between the terms *incentive* and *reward*. An incentive is "something that incites a tendency to determination or action," whereas a reward is "something that is given in return for good done" (Webster, 1971). Because the difference between these terms tends to be obscured, I will use them interchangeably in what follows.

Background

Up until World War I, employee income was largely determined by straight-time pay for hours worked (Wiatrowski, 1990). Since then, a variety of social, political, and economic factors have contributed to the growth of benefits so that health and life insurance, retirement income, and vacation time, for example, became standard features of compensation packages. Workers have come to expect them and, increasingly, labor disputes have centered on benefit concerns, usually those related to health care (Dauer, 1989).

As the century progressed, the demographic profile and needs of Americans shifted and benefit packages became more comprehensive. But as plans changed, so did their costs. Benefits accounted for 17% of compensation in 1966, but rose to nearly 28% by 1990 (Bureau of Labor Statistics, 1977, 1990). Likewise, health care costs increased dramatically. As a share of total compensation, business spending for health care grew from 2% in 1965 to 4.9% in 1980 to 7.1% in 1990 (Levit & Cowan, 1991). At the start of the decade, health care costs represented 45.5% of all benefits, compared to 22.4% in 1965 (Levit & Cowan, 1991).

Many of these cost increases can be traced to the way benefits, and in particular health benefits, have been designed and administered. Since 1960 and continuing into at least the early 1980s, health care was largely subsidized by the employer with few restrictions to limit employees and their providers from using and perhaps abusing the use of medical care. Workers typically argued for and received health care protection that offered full coverage of medical expenses without regard for employee contributions (Wiatrowski, 1990). However, as subsequent research by the Rand Corporation indicated, as the support for health care increased, the overall demand for and total cost of health care also increased (Keeler & Rolph, 1983). Ironically, as support increased—even the provision for totally free care—the status of health did not necessarily change (Brook, Ware, Rogers, et al., 1983).

The 1980s represented an era of growing concern for organizational health care costs. Total business expenses for private health insurance rose from approximately 48 billion dollars at the start of the decade to nearly 140 billion dollars by 1990 (Levit & Cowan, 1991). This increase resulted in the present-day movement to control costs and the adoption of such strategies as managed care. In the process, companies adopted a "take-away" mentality and added numerous administrative controls (Jose, 1991). For example, while increasing out-of-pocket expenses, typical program features included limitations on providers, prehospitalization certification, mandatory second opinions, and concurrent utilization review. But despite these efforts, health care costs continued to rise (Vincenzino, 1989), and employers now found health care at the forefront of major labor disputes (Employee Benefit Plan Review, 1989).

At the same time, growing research indicated that medical costs were not distributed randomly among an employee population. Genetics and environmental factors notwithstanding, health care costs were far from randomly distributed. Marked differences were shown to exist between individuals on the use of health care based on lifestyle, or behaviors that the individual presumably chose to adopt freely. For example, the Control Data Corporation found a positive relationship between the level of employee behavior risk and health care costs. As the level of risk increased so did monthly claims, hospital inpatient days, and percentage of high claimants (Milliman & Robertson, Inc. & Control Data Corp., 1987). A small but growing literature came to the same conclusion—workers with negative health behaviors or behaviorally based biomedical measures (weight, blood pressure) cost more in terms of health care than their exercising, nonsmoking and seat belt–wearing counterparts (Yen, Edington, & Witting, 1991; Vickery, Golaszewski, Wright, & McFee, 1986).

The same conclusion was supported for the use of time off from work through the use of sick leave. Similar to health care costs, research also indicated differences in the use of sick leave based on lifestyle. The Du Pont Company reported that employees with any of six behavioral risks have between 10% and 39% higher levels of absenteeism than employees without these risks (Bertera, 1991).

When other health-related factors are considered in this analogy, such as disability compensation, the differences in benefit costs between individuals who practice prudent lifestyles and those who do not may be enormous. Du Pont's research conservatively estimates that the cost of behaviorally related illnesses to the company is $70.8 million per year (Bertera, 1991).

In summary, several major lessons were learned during the past decade as the cost of health care skyrocketed:

- Personal employee behavior has a far greater consequence on health care costs than previously realized.
- The provision for employer-provided health insurance, no matter how costly, does not necessarily guarantee good health.
- Attempts to cut benefit costs by "take away" mechanisms only antagonized workers while achieving negligible effects on the rate of health care spending.

The Case for Incentives: Organizations Get What They Pay For

A substantial literature supports the contention that whatever behaviors an organization chooses to reward will increase, while those that it chooses to punish or, more appropriately, ignore, will ultimately decrease (Berry & Houston, 1993). As the field of Organization Development or the science of creating change in organizations indicates, one of the primary influences supporting an organization's culture, and ultimately how people behave in it, is what it chooses to reward (Beer, 1980). In retrospect, despite noble intentions, organizations may have created their own health care cost crisis by inadvertently rewarding poor health behaviors with liberal health insurance, sick leave, and other benefits at the expense of undermining good health behavior (Egger, 1980).

This focus is starting to change, partly because benefits are no longer so generous, but also because companies are taking a more enlightened view of their need to support healthy behavior. Today 81% of worksites of 50 or more employees provide some type of health promotion activity (Office of Disease Prevention and Health Promotion, 1992). However, despite periodic reports of success (Kaman, 1987), managers question health promotion's ability to contain health care costs (*Business and Health*, 1990). In part, this skepticism persists because widespread adoption of healthy behaviors by employees and their dependents remains elusive (Allen & Kraft, 1984).

On the other hand, recent thinking in program implementation indicates that for significant numbers of employees to adopt healthy behaviors, the underlying culture of the organization must change (Goodman & Steckler, 1990). Borrowing again from Organization Development (Beer, 1980), in order for worksite cultures to change and support a new behavior

1. individuals must be educated accordingly,
2. desired behaviors must be demonstrated through role modeling,
3. selection of new employees should be based, in part, on their ability to demonstrate the desired behaviors, and
4. the behaviors in question must be rewarded (Beer, 1980).

If the above is true, then rewards and how they are administered are a necessary, though not sufficient, factor in the widespread adoption of healthy behaviors by employees.

Recent discussions on the role of health promotion in work organizations support this contention. A report from the Pew Fellows Conference—a group of 26 senior managers responsible for corporate health affairs—concludes that programs need to be multifaceted and long-term and address the underlying attitudes, values, and beliefs of employees, their social supports, and their economic interests—not just their risk factors (Chapman-Walsh & Egdahl, 1989). This approach requires changes in policy, social and group norms and values, and the way organizations provide financial incentives.

Incentives in Worksite Health Promotion

The use of incentives is a well-established practice in the worksite health promotion field (Shepard & Pearlman, 1982). Incentives by their broadest definition include the use of group competitions, lotteries, and direct financial payments for behavioral outcomes (Warner, 1990). However, for the purposes of this discussion, the focus of examples will be limited to those that offer more tangible and financially relevant rewards. The following section reviews incentives generically and how they have been applied in a variety of contexts.

One of the earlier examples of such a mechanism was the use of direct payments by the Bonnie Bell Cosmetics Company to employees who demonstrated physical exercise behavior (Feldman, 1985). Employees received one dollar for every mile run, 50 cents for each mile walked, and 25 cents for each mile on the bicycle. Though influencing employee behavior, this approach proved "too successful" because the firm could not afford the expense and had to discontinue its use.

Bonnie Bell has since reinstituted many of its earlier program features and added a number of other components such as extended lunch hours to exercise and discounts on athletic equipment (Chapman, 1991). But this practice established a pattern—employees *do* respond to rewards.

Davidson Louisiana, Inc., a Louisiana based building products distributor, conducted a unique employee fitness program whereby employees could earn points toward an expense paid vacation to cities around the world, for example to Rome or Mexico City (Brown, 1981). A previous incentive system involved traditional business criteria such as reaching sales quotas or attending a set number of sales meetings. Davidson added health and fitness criteria to this list. Employees could now earn points for meeting prescribed exercise regimens and for passing age- and sex-adjusted biomedical tests (weight, blood pressure, fitness). According to reports, about 80% of Davidson's eligible employees chose to participate in the program and, of those, nearly 80% qualified for the vacation (which also required reaching at least two other sales goals). What made this program unique was the fact that Davidson had no fitness facility. All employees exercised on their own time and made use of whatever resources they had available.

Using a similar model, the Westinghouse Corporation implemented a pilot program integrating health promotion with a variable compensation plan at its College Station, Texas, facility (Benefits Today, 1992). A bonus system was started providing financial incentives for meeting traditional workplace goals. Added to this list was a provision for use of the company's fitness center. A year-end bonus of $200 was made available to employees for completing a minimum of 10 minutes of center-based aerobic activity for at least 60% of available workdays. Simply walking briskly for 10 minutes could apply.

Results indicated that 97% of employees were satisfied with the program and that an impressive 51% of all employees used the center regularly. Of those eligible, 58% met the reward criterion.

Other novel incentive systems that have been used include the following. The Speedcall Corporation reported a reduction in the prevalence of smoking among its workers from 67% to 43% within a month following the implementation of a $7 weekly bonus to the paychecks of nonsmokers (Shepard & Pearlman, 1982). Forster and colleagues (1985) found that a worksite weight management program using a small financial reward for weight loss was effective in motivating employees to lose weight. Oswald (1989) reported that workers who received free low-fat lunches decreased their mean cholesterol levels nearly twice as much as employees who did not.

However, despite promising results, incentives do not guarantee success. A number of studies reveal inconclusive findings with the use of incentives for smoking cessation (Matson, Lee, & Hopp, 1993), weight loss (Kramer, Jeffery, Snell, & Forster, 1986), and seat belt-use programs (Mullen, 1988). As Warner (1990) indicates, incentive systems fail to teach fundamental principles of behavior change. High rates of initial change tend not to be sustained over time, especially when incentives cease.

Warner's comments are valid. Behavior change results from a confluence of factors, incentives being only one. But insights from the study of motivation may also provide answers. Expectancy theory indicates that motivation is a function of both the belief that a behavior will result in a reward and that the reward is of meaningful value (Lawler, 1973). For an employee to be motivated by an incentive, two expectations must occur: He or she must believe that a reward will be provided if an outcome is realized, and the reward must be perceived to have sufficient value to warrant the effort. More recently, the element of self-efficacy has been added to motivation theory (Strecher, McEvoy, Becker, & Rosenstock,

1986). Self-efficacy is the belief that one has the capability of performing a behavior—in other words, the behavior is perceived as achievable.

But how does this relate to benefit designs? First, by being included in an established benefits package and linked to company policy, employees perceive incentives as both viable and ongoing. Second, benefits or direct cash payments connected to them have real value that most employees can acknowledge. And, third, as the following section of this chapter indicates, program managers have developed multiple and varied approaches in their administration that allow employees to perceive success.

Using Benefits as Health Behavior Incentives

The use of benefit-related incentives was first examined in the 1992 National Survey of Worksite Health Promotion (ODPHP, 1992). In companies of 50 or more employees, 31% offered flexible spending accounts for health care and health promotion expenses, 8% provided annual fixed amount reimbursement for health promotion services, 12% risk rated on smoking status, and 13% subsidized discounts or reduced fees for fitness facilities. The remainder of this chapter reviews examples of these types of programs.

To identify suitable models for this discussion, a computer search of the literature was undertaken, and numerous conversations were held with program managers, academicians, and consultants in the worksite health field. Many excellent examples were identified; however, in the interest of brevity, only a sampling has been provided. Their selection was based in large part on their comprehensiveness, adherence to accepted behavior change theory, and their adequacy of history to draw conclusions.

The Adolph Coors Company

The Coors Brewing Company of Golden, Colorado, emphasizes a broad-based approach to health care cost management that includes a variety of administrative controls in combination with the promotion of employee health. One feature of its effort is the provision for financial incentives tied to the benefits plan. For example, nonsmokers pay 50% less for supplemental life insurance than smokers. Employees are also provided additional life insurance if they or their spouses are killed in automobile accidents while wearing seat belts (Gilfillan, 1992). However, the most notable component of the Coors plan is the provision for risk-rating of health insurance (Kaelin, Barr, Golaszewski, & Warshaw, 1992). Coors will provide a 5% reduction for insurance copayments if employees meet predetermined levels of risk as identified by health risk appraisal (HRA). The HRA consists of a questionnaire that determines the employee's statistical "health age" based on the presence or absence of certain risk factors, including clinical measures. Individuals within an appraisal-based health age standard qualify for the award.

Of approximately 20,000 Coors employees or spouses completing the HRA, nearly 75% met the award criteria. Although 15% were found to be at risk, approximately 50% of these reduced their risks within the next 12 months to later qualify.

Coors has extensively evaluated its health promotion effort and concluded that the program returned between $1.24 and $8.33 for every dollar invested (Terborg, 1988). Many factors may have accounted for this finding, and subsequently, the exact role of incentives is not known. However, Coors notes that participation in onsite wellness programs increased by 40% since the HRA assessment and award system were introduced (Gilfillan, 1992).

Southern California Edison

In 1989, the Southern California Edison (SCE) Company of Rosemead, California, introduced its Health Flex program as one of a number of initiatives to curb its rapidly increasing cost of health care (Ham, 1989). One component of the program, the Good Health Rebate, offers employees and their spouses covered under their flexible benefits plan rebates of $240 annually on their premium contributions. Similar to the Coors plan, rebates are based on passing five modifiable risk factors within acceptable levels: body weight, blood pressure, smoking, cholesterol, and blood sugar levels (Kaelin et al., 1992). Those who do not pass may appeal the original medical findings and submit to further testing, or they may qualify by either seeking medical treatment or completing an appropriate educational program. To encourage participation, screening costs can be deducted from the company's Preventive Health Account, a fund to support preventive medical services up to a maximum of $500 per person per year.

A second feature is the availability of the Good Health Account. SCE offers a $150 credit to both employees and their spouses that can be applied to the cost of prevention and/or wellness services (e.g., membership in a fitness center). Other preventive services such as mammography and Pap smears are covered under the Preventive Health Account.

Results indicate that participation in screening ranges between 36% and 48% of eligible employees. The program appears to reach high-risk individuals, as over 51% of the highest quartile of claimants are actively involved, and large numbers of employees with previously unknown risk factors have been identified. Although evaluation remains ongoing, trends indicate decreasing risks for smoking, excessive weight, and hypertension (telephone conversation, M. Schmitz, Southern California Edison, March 1993).

The Lord Corporation

The Lord Corporation, an industrial and chemical manufacturer located in Erie, Pennsylvania, takes a somewhat different approach to health care cost containment by emphasizing medical consumer education. Lord's "Cost Containment Program" is a mixture of seminars, literature, and cost-comparison data, supplemented by policies that help workers make more cost-effective medical care decisions (Health Action Managers, 1990).

Several policy innovations in the Lord program apply to the benefits-incentive model. First, employees are provided resources to make them better medical consumers, including an updated medical consumer guide, a medical self-care text, and ongoing information about waste in the health care system in "LordFacts," the company newspaper. Second, pricing information provided on local hospitals is supplemented by comparative cost data from preferred providers located within a 2-hour drive of Erie. Substantial savings for the same treatment can be realized by using these out-of-town services. To encourage use, Lord pays all travel expenses for both the employee and a companion. However, the real incentive feature of the program is its provision for "gain-sharing" of health care cost savings. If actual health care costs are less than the annual budgeted amount (Lord is self-insured), the savings are divided equally between employees and the company.

Lord does not limit its efforts to medical consumer education. The company also provides a number of wellness programs. At larger corporate facilities, employee wellness committees sponsor such activities as walking and running clubs and special events such as cross-country ski outings and weight loss contests. To further stimulate activity, the company also reimburses employees for offsite wellness activities that include commercial fitness center memberships and smoking cessation classes.

In 1993, Lord introduced "Life Share" as a new component of its program. In Life Share, employees are rewarded points for meeting or completing a wide variety of health behaviors, biomedical standards, and screening activities. Earned points convert to dollars that are applied to the health insurance premium. Any remaining monies revert back to the employee as a cash bonus.

Although Lord's program has not been systematically evaluated, employees did receive dividends during 3 of the first 6 years of operation. The last 2 years have not been as promising as health care spending surpassed budgeted amounts. However, company officials indicated that unrealistically low budgets may have accounted for this finding. Nevertheless, Lord is optimistic about the impact of its cost containment efforts. Since its inception 8 years ago, health care cost increases have averaged about 12% per year, a figure considerably lower than national averages (telephone conversation, R. Shindledecker, Lord Corporation, March 1993).

Quaker Oats

Quaker Oats was experiencing rapidly escalating health care costs of between 20% and 30% annually during the early 1980s. In response to this problem, the company implemented an ambitious redesign of its benefits plan to reduce health care expenses without having to shift costs to employees in the process (Penzkover, 1984).

Quaker's "Health Incentive Plan" consisted of four components: (1) a typical comprehensive medical plan, (2) an annual economic adjustment to the benefits plan reflecting changes in inflation, (3) an employee health care expense account that reimbursed medical costs to a predetermined maximum, and (4) an incentive plan that would provide employees with a cash payment if total health care spending was below an annual targeted amount.

To enhance the incentive component, two basic strategies were employed. First, Quaker Oats initiated a medical consumer program called "Informed Choice." Employees were given copies of a medical self-care book and an internally developed health care consumer's guide. In addition, the company published a pricing guide for local hospitals, which listed charges for selected services at approximately 100 competing hospitals in the Quaker Oats operations area. Taken in total, this material provided information that allowed employees to make more informed choices as health care buyers.

Second, Quaker Oats adopted the wellness program "Live Well–Be Well." The program consisted of screening and health risk appraisal, a health resource center, and a number of group and self-paced learning sessions on various health topics.

Quaker Oats has since added to its wellness effort with the adoption of "QuakerFlex." Employees can now earn up to $250 for themselves and $250 more for their spouses by earning QuakerFlex Healthy Lifestyle credits. Credits are earned by taking a health risk appraisal, including screening activities, and signing a pledge to maintain or begin healthy behaviors (e.g., seat belt use or aerobic exercise). Credits can be used in the Quaker benefits package to purchase additional health coverage, buy additional vacation time, be deposited in the company's 401K plan, or be taken as cash. QuakerFlex has just begun in 1993, but preliminary results indicate that the program is attracting more individuals to screening activities, including spouses.

The company reported that in the first 2 years of operation, medical claims dropped from $1,421 to $1,336 to $1,324 per employee, per year, respectively. Employees received a dividend in each year, as total plan costs (including cash refunds) did not exceed the annual targeted amount (Penzkover, 1988). In contrast to the 20% annual increases experienced during the 10 preceding years, costs increased an average of only 5.2% per year. Quaker Oats has not published findings since, but according to company officials they have continued to experience significantly lower health care costs than other companies nationally (telephone conversation, C. Kahn, Quaker Oats Company, February 1993).

Foldcraft Company

The previous examples represented companies of substantial size. However, companies need not be large to offer innovative benefit packages. The Foldcraft Company, a 300-employee manufacturing firm in Kenyon, Minnesota, is one such example (American Health Consultants, 1990).

With increasing health care costs during the 1980s, large numbers of Foldcraft employees were dropping out of the company's health care plan. To lure employees back to the plan (insufficient numbers were enrolling to negotiate favorable contracts) and improve employee health, Foldcraft first attempted to provide a $15 to $20 monthly bonus to individuals who participated in fitness activities as a way to offset insurance costs. However, this approach rewarded only those employees already active and had little impact on attracting individuals to the health care plan. Subsequently, Foldcraft adopted an ambitious health screening and risk-rated wellness program.

First, if employees and their spouses completed a screening and health risk appraisal, they received a $20 per month rebate on their health insurance premium. To encourage participation, 1.5 hours of paid time off were given to complete data collection and have the HRA results interpreted. Second, up to $40 of additional health insurance rebates were offered per month depending on the level of individual risk. Rebates were tied into achieving acceptable levels for smoking, percentage of body fat, blood pressure, cholesterol, oxygen uptake, flexibility, and later, abdominal strength. An overall score was calculated and levels of risk were determined and translated into rebates (which could amount to $720 a year). A later provision added a half day of vacation time for employees completing a voluntary drug screening.

No formal evaluation of this effort has been undertaken, but a number of positive observations were made. After only 2 months, an additional 10% of employees returned to the health care plan, and a 15% increase in the number of new employees joining the plan was observed. Eighty-two percent of employees participated in the health screening, and nearly 90% participated in the drug screening. Attendance at weekly wellness meetings increased and more positive eating patterns were observed at daily coffee breaks (e.g., fewer boxes of donuts were being delivered).

Although insurance enrollments have remained steady, company officials indicate that there is a much greater employee awareness of the impact of lifestyle on health. Overall, employees appear to accept the program well. More people are exercising, and the use of company provided wellness programs is high. Of note, Foldcraft is a 100% employee-owned company. Employees have a huge personal stake in controlling health care costs and preserving the economic well-being of the organization (telephone conversation, D. Erickson, Foldcraft Company, March 1993).

Other Approaches to Incentive-Related Benefit Plans

A number of other examples have been cited for their innovative approaches in using benefits as incentives. In brief review, the city of Birmingham, Alabama, mandates an annual screening and health risk assessment in order for employees to enroll in the health insurance plan (Harvey et al., 1993). Dominion Resources, Inc., of Richmond, Virginia, pays its workers an extra $10 per month in pretax flexible benefits credits if they qualify as low-risk in five lifestyle-related categories (Woolsey, 1992). And Union Camp Corporation headquartered in Wayne, New Jersey, adds a $100 Healthy Lifestyle Award to the company's health care spending account if employees self-certify they are nontobacco users (Kaelin et al., 1992).

Other novel benefit-related incentives that have been proposed include additional sick leave allowances, disability waiting period reductions, elevated disability payments for meeting predetermined health risk criteria, earned "well days," and carry-forward and/or cash-out provisions for unused sick leave (Chapman, 1991).

Summary and Conclusions

As this discussion indicates, a movement by employers to provide financially relevant incentives to stimulate healthy behavior is under way. Increasingly, many of these initiatives are benefit-plan related. Based on the previous review, the following is noted:

1. Trends or anecdotal findings suggest that benefit incentives stimulate the use of health promotion services and/or influence the adoption of healthy behavior.
2. Employees appear to respond favorably to the implementation of these plans.
3. When benefit incentives are used in conjunction with traditional cost-management strategies and health promotion services, health care costs appear to be better managed.
4. Management recognizes the value of these approaches and movements are under way toward greater program expansion.

Even without the benefit of an empirical research base, a strong argument can be made for the use of benefits as incentives. This assertion is especially true when the results of health promotion programs are combined with those of numerous studies found in the psychology literature that use similar models to influence employee behavior. Therefore, this chapter endorses the use of benefit incentives as a health promotion strategy and, with some qualification, encourages organizations to incorporate similar benefit features in their operational plans. Implementation strategies are readily available elsewhere (Chapman, 1991; Golaszewski, Kaelin, Miller, & Douma, 1992).

In the interest of objectivity, however, it should be noted that not all experts support the use of benefit-related incentives. The use of risk-rating in particular draws considerable criticism. Experts argue that certain risks are a function of genetics (e.g., cholesterol and hypertension), and individuals should not be penalized for what they cannot control (Terry, 1991). Furthermore, simply determining risk can be a problem. Error in measurement is a concern, but beyond testing accuracy, the exact level of what constitutes a risk is not always clear. Others argue that risk-rating defeats the historical purpose of insurance by shifting costs to a few rather than distributing them over the wider population (Kaelin et al., 1992). Risk-rating may also suffer from legal challenges from the American Disabilities Act, which prohibits discrimination to individuals based on handicaps (Priester, 1992). Depending on the law's interpretation, handicaps might include genetically based obesity or tobacco addiction. Aside from risk-rating, benefit-related incentives are not empirically tested—their true effectiveness remains unknown. As a result, monies spent for this purpose might better be used for improving educational programs or communicating their availability to employees. Finally, without careful planning, incentives could actually produce unexpected and undesirable side-effects (Chapman, 1991). For example, "Wellness Day" bonuses for not using sick leave could encourage ill employees to come to work and spread infection to others.

As the preceding indicates, new approaches in health promotion delivery create a profusion of questions on their efficacy and impact. Accordingly, a series of research questions was recently posed for the area of risk-rated health insurance (Kaelin et al., 1992). These same questions apply equally to the broader area of benefit-related incentives, and include the following:

1. Do benefit incentives increase the use of health promotion services? With or without an increase in services, do they improve health behavior, decrease morbidity, and affect health care costs, absenteeism, and other economic variables?
2. What are the effects of benefit incentives on employee morale and satisfaction with the benefits plan?
3. Do benefit incentives help or hinder employee recruitment and/or retention? Do they enhance the recruitment of health-oriented employees?
4. Are there grounds for legal challenges to varying benefits distribution based on health criteria?

The worksite health promotion movement has gone through a series of changes that shaped the discipline into its current status as an instrument of national policy. Strategies dating to the 19th century emphasized information dissemination with an accent on print media and classroom-based programs. The 1970s gave rise to the fitness center boom and hundreds of companies constructed state-of-the-art facilities that rivaled commercial centers for comprehensiveness and allure. During the late 1970s and early 1980s, programs began to adopt technology and the use of computers, telephones, and video for educational purposes grew rapidly. More recently, organizations recognized the need to support positive health behavior through a strategy of shaping organizational values and norms. In this context, incentives, especially those related to benefits, helped to reinforce the practice of healthy behavior.

In summary, exciting innovations in benefit designs are emerging, and trends indicate that this practice will grow. Although hard data are lacking, the available evidence to support this movement is encouraging. However, the real enthusiasm for this effort is not so much for the lessons that have been learned, but rather for the realization that the most effective models have yet to be developed.

Acknowledgments

The following individuals are recognized for their support and contributions: Darwin Dennison, EdD, Kathryn Dennison, EdD, Dominic Galante, MD, and Cynthia Jagodzinski, DINE Systems, Inc.; Christopher Blodgett, PhD and Robert Delprino, PhD, SUNY College at Buffalo; Mark Kaelin, EdD, Montclair St. College; Larry Chapman, MPH, Corporate Health Designs; David Chenoweth, PhD, East Carolina University; and William Jose, PhD, Northwestern College of Chiropractics.

References

Allen, R., & Kraft, C. (1984). The importance of cultural variables in program design. In M. O'Donnell and T. Ainsworth (Eds.), *Health promotion in the workplace* (pp. 63-96). New York: Wiley & Sons.

American Health Consultants (1990). Linking health promotion to benefits: Small company pleased with wellness incentive plan. *Employee Health & Fitness*, **12**, 100-107.

Beer, M. (1980). *Organization change and development: A systems view*. Santa Monica, CA: Goodyear.

Benefits Today (1992). Westinghouse adds financial incentive to fitness program to boost participation. *Benefits Today*, **9**, 263-264.

Berry, L., & Houston, J. (1993). Improving work performance: The motivation to work. In L. Berry & J. Houston (Eds.), *Psychology at work* (pp. 75-111). Madison, WI: Brown and Benchmark.

Bertera, R. (1991). The effects of behavioral risks on absenteeism and health-care costs in the workplace. *Journal of Occupational Medicine*, **33**, 1119-1124.

Brook, R., Ware, J., Rogers, W., Keeler, E., Davies, A., Donald, C., Goldberg, G., Lohr, K., Masthay, M., & Newhouse, J. (1983). Does free care improve adults health? *New England Journal of Medicine*, **309**, 1426-1434.

Brown, J. (1981). An incentive-based employee fitness program. *Health Education*, **12**, 23-24.

Bureau of Labor Statistics (1977). Employee compensation in the private nonfarm economy, 1974 (Bulletin 1963). Washington, DC: Department of Labor.

Bureau of Labor Statistics (1990). *Employment cost indexes and levels, 1975-90* (Bulletin 2372). Washington, DC: Department of Labor.

Business and Health (1990). The 1990 national executive poll on health care costs and benefits. *Business and Health*, **8**, 25-38.

Chapman, L. (1991). *Using wellness incentives: A positive tool for health behavior changes*. Seattle: Corporate Health Designs.

Chapman-Walsh, D., & Egdahl, R. (1989). Corporate perspectives on work site wellness programs: A report on the seventh Pew Fellows Conference. *Journal of Occupational Medicine*, **31**, 551-556.

Chenoweth, D. (1988). *Health care cost management*. Indianapolis, IN: Benchmark Press.

Dauer, C. (1989, October). Unions focus on health care. *National Underwriter*, pp. 63-64.

Egger, R. (1980, April). Let's pay patients to stay well. *Medical Economics*, pp. 123-128.

Employee Benefit Research Institute (1987). *Fundamentals of employee benefit programs*. Washington, DC: Author.

Employee Benefit Plan Review (1989, December). *Health care issues dominate contract talks*.

Feldman, R. (1985). The assessment and enhancement of health compliance in the workplace. In R. Feldman and G. Everly (Eds.), *Occupational Health Promotion: Health Behavior in the Workplace* (pp. 33-46). New York: Wiley & Sons.

Forster, J., Jeffery, R., Sullivan, S., & Snell, M. (1985). A work-site weight control program using financial incentives collected through payroll deduction. *Journal of Occupational Medicine*, **27**, 804-808.

Gilfillan, L. (1992). Using an HRA in a corporate health promotion program. In K. Peterson (Ed.), *Directory of Health Risk Appraisals* (pp. 39-44). Indianapolis, IN: Society of Prospective Medicine.

Golaszewski, T., Kaelin, M., Miller, R., & Douma, A. (1992). Combining health education, risk-rated insurance and employee rebates into an integrated health care cost management strategy. *Benefits Quarterly*, **8**, 41-50.

Goodman, R., & Steckler, A. (1990). Mobilizing organizations for health enhancement: Theories of organizational change. In K. Glanz, F. Lewis, & B. Rimer (Eds.), *Health Behavior and Health Education* (pp. 314-341). San Francisco: Jossey-Bass.

Green, L., & Johnson, K. (1983). Health education and health promotion. In *Handbook of Health, Healthcare, and the Health Professions* (pp. 744-765). New York: The Free Press.

Ham, F. (1989). How companies are making wellness a family affair. *Business and Health*, **7**, 27-33.

Harvey, M., Whitmer, R., Hilyer, J., & Brown, K. (1993). The impact of a comprehensive medical benefit cost management program for the city of Birmingham: Results at five years. *American Journal of Health Promotion*, **7**, 296-303.

Health Action Managers (1990). The buck stops here. *Health Action Managers*, **4**, 1-10.

Jose, W. (1991). Financial incentives for healthy lifestyles: Accountability and equity. In S. Muchnick-Baku (Ed.), *The Challenge of Financial Incentives and Risk Rating* (pp. 21-24). Washington, DC: Washington Business Group on Health.

Kaelin, M., Barr, J., Golaszewski, T., & Warshaw, L. (1992). Risk-rated health insurance programs: A review of designs and important issues. *American Journal of Health Promotion*, **7**, 118-128.

Kaman, R. (1987). Cost and benefits of corporate health promotion. *Fitness in Business*, **2**, 39-44.

Keeler, E., & Rolph, J. (1983). How cost sharing reduced medical spending of participants in the health insurance experiment. *Journal of the American Medical Association*, **249**, 2220-2222.

Kramer, F., Jeffery, R., Snell, M., & Forster, J. (1986). Maintenance of successful weight loss over 1 year: Effects of financial contracts for weight maintenance or participation in skills training. *Behavioral Therapy*, **17**, 295-301.

Lawler, E. (1973). *Motivation in work organizations*. Monterey, CA: Brooks/Cole.

Levit, K., & Cowan, C. (1991). Business, household, and governments: Health care costs, 1990. *Health Care Financing Review*, **13**, 83-93.

Matson, D., Lee, J., & Hopp, J. (1993). The impact of incentives and competitions on participation and quit rates in worksite smoking programs. *American Journal of Health Promotion*, **7**, 270-287.

Milliman & Robertson, Inc., & Control Data Corp. (1987). *Health risks and behavior: The impact on medical costs*. Brookfield, WI: Milliman and Robertson.

Mullen, P. (1988). Health promotion and patient education benefits for employees. *Annual Review of Public Health*, **9**, 305-332.

Office of Disease Prevention and Health Promotion (1992). *1992 national survey of worksite health promotion activities*. Washington, DC: Public Health Service.

Oswald, S. (1989). Changing employees' dietary and exercise practices: An experimental study in a small company. *Journal of Occupational Medicine*, **31**, 90-97.

Penzkover, R. (1984). Building a better benefit plan at Quaker Oats. *Business and Health*, **2**, 33-36.

Penzkover, R., (1988). Health incentives help clamp costs. *Personnel Journal*, **68**, 114-118.

Priester, R. (1992). Are financial incentives for wellness fair? *Employee Benefits Journal*, **17**, 38-40.

Shepard, D., & Pearlman, L. (1982). *Incentives for health promotion in the workplace: A review of programs and their results*. Boston: Center for Analysis of Health Practices.

Strecher, V., McEvoy, B., Becker, M., & Rosenstock, I. (1986). The role of self-efficacy in achieving health behavior change. *Health Education Quarterly*, **13**, 73-91.

Terborg, J. (1988). *Cost benefit analysis of the Coors Wellness Program*. Golden, CO: The Coors Brewing Company.

Terry, P. (1991, February 25). A dangerous innovation. *Health Action Managers*, pp. 1, 8-9.

Vickery, D., Golaszewski, T., Wright, L., & McFee, L. (1986). Lifestyle and organizational health insurance costs. *Journal of Occupational Medicine*, **11**, 1165-1168.

Vincenzino, J. (1989). Trends in medical care costs—update. *Statistical Bulletin of the Metropolitan Insurance Company*, **70**, 26-34.

Warner, K. (1990). Wellness at the worksite. *Health Affairs*, **9**, 64-79.

Webster's Seventh New Collegiate Dictionary (1971). Springfield, MA: G.C. Merriam.

Wiatrowski, W. (1990, March). Family-related benefits in the workplace. *Monthly Labor Review*, pp. 28-33.

Woosley, C. (1992, February 17). Varied paths lead to common goal of wellness. *Business Insurance*, pp. 16-18.

Yen, L., Edington, D., & Witting, P. (1991). Associations between employee health-related measures and prospective medical claims costs in a manufacturing company. *American Journal of Health Promotion*, **6**, 46-54.

Chapter 11

Computer Simulation: A Promising Technique for the Evaluation of Health Promotion Programs at the Worksite

James R. Terborg

Research has documented the benefits of worksite-based health promotion programs on employee behavior and health risk status (Blair, Piserchia, Wilbur, & Crowder, 1986; Fielding, 1984; Glasgow & Terborg, 1988). Data further suggest that such programs can be cost-effective compared to nonworksite programs (Erfurt, Foote, & Heirich, 1992; Opatz, 1987; Opatz, Chenoweth, & Kaman, 1991). Less is known about economic benefits to companies with health promotion programs that may result from presumed reductions in health care costs, absenteeism, accidents, and turnover and improvements in employee morale and productivity (Shephard, 1992; Warner, Wickizer, Wolfe, Schildroth, & Samuelson, 1988). Although statistically reliable associations have been reported between employee health behavior and organizational outcomes (e.g., Bertera, 1990; Bernacki & Baun, 1984), considerable uncertainty exists regarding the cost-benefits of worksite health promotion programs.

The primary reasons for this uncertainty are the paucity of rigorous experimental designs and difficulty in acquisition of even basic data (Shephard, 1992; Warner et al., 1988). Although executives and managers often ask for documentation of benefits, the 1992 national survey of worksite health promotion activities (U.S. Department of Health and Human Services Public Health Service, 1993) showed that formal evaluations of program effects were conducted by only 12% of those responding. Furthermore, in the few companies doing formal evaluations, data on outcomes such as health care costs, employee morale, health status, absenteeism, and productivity were often not available for analysis.

The importance of experimentation is widely recognized, but advances in knowledge are limited by constraints placed on researchers operating in the worksite. Even if long-term access to a sample of worksites were obtained and worksites were randomly assigned to treatment conditions, the costs of evaluation would be substantial. Companies wishing to make a specific evaluation of their worksite program might find that the cost of such analyses would exceed the presumed savings. Furthermore, even if the results of such studies were available the results might not generalize to other worksites with different employee demographics, medical plans, and turnover rates.

One promising solution to this problem is the development and use of computer simulations that allow for a detailed examination of the presumed benefits attributed to health promotion programs at the worksite. The next section will briefly introduce the content and methodology of computer simulations.

What Are Computer Simulations?

A computer simulation is nothing more than a model of a system or process that is accessed with a computer (Whicker & Sigelman, 1991). By conducting experiments with the model it is possible to begin to understand how the system or process functions and to evaluate various strategies for its operation.

To help demystify computer simulations, consider the task of designing a house. When an architect prepares blueprints of the house, for example, the architect is creating a simulation, or model, of what the house will look like when it is constructed. Once the model has been developed, it is relatively easy to explore the impact of proposed changes to the building on such things as overall appearance or cost. For example, the architect might want to experiment with the use of gables in the roof to see how that affects appearance, usable floorspace, and cost of construction.

Recently, computer programs have been written that allow architects to design on the computer instead of at the drafting table. By using computer programs to develop blueprints for a house and to experiment with different design features, the architect is conducting a computer simulation.

Computer simulations can take many forms to model many different phenomena. In the example listed above, a physical model of a physical system (a house) was created. But computer simulations can also be symbolic models of complex and dynamic processes. Stoll (1983) developed a computer model to study why nations to go war with one another, and Lewandowsky and Murdock (1989) used computer simulations to test theories of human learning and memory.

Computer simulations are just beginning to be applied to the field of health care and promotion. Pallin and Kittell (1992) developed a model of patient flow through an emergency room ward. They then experimented with different procedures for assigning patients to treatment. The computer simulation showed that one procedure, called "fast track," could produce a 50% reduction in staff and resources with no deterioration in level of patient care for one type of patient group. Hatcher and Rao (1988) developed a computer model to evaluate a proposed Health Promotion Center in a private hospital. The purpose of the simulation was to examine the impact of different pricing, advertising, and service strategies on profitability.

Having provided a few examples of how computer simulations can be used, it is time to go into more detail about the content of computer simulations. Computer simulations have several components: the assumptions that underlie the model; the independent, dependent, and control variables included in the model; and the specific rules, called algorithms, that determine relationships among these variables (Whicker & Sigelman, 1991). Computer simulations share many features typically associated with standard inductive research methods in that both employ independent, dependent, and control variables. The major difference is that in a simulation the values taken by the dependent variables are entirely determined by the algorithms driving the simulation.

Computer simulations offer several advantages over standard empirical research, but they are not substitutes for empirical research. Compared to standard research designs, simulations are flexible, efficient, and able to handle many variables. Simulations are especially useful when the problem under consideration is one about which empirical research is costly, difficult, unusually complex, or unethical but where a basic knowledge of how the system operates is known. Having some understanding of how the system operates is critical because without this it is almost impossible to define algorithms that accurately depict how variables in the system relate to each other.

Computer Simulations and the Evaluation of Worksite Health Promotion Programs

Given the current state of knowledge regarding the economic consequences of worksite-based health promotion programs, the development and use of computer simulations is highly appropriate. Some knowledge already exists on the costs and benefits associated with health promotion programs (for examples, see Bertera, 1991; Erfurt, Foote, & Heirich, 1992; Golaszewski, Snow, Lynch, Yen, & Solomita, 1992; Opatz, 1987; Opatz, Chenoweth, & Kaman, 1991; Wood, Olmstead, & Craig, 1989; Yen, Edington, & Witting, 1991, 1992). These data can be used as starting points in the development and use of computer simulations as a technique for evaluating the economic impact of health promotion programs at the worksite.

Some work already has been done that makes projections of costs or savings using regression analyses and other inferential statistical techniques, although these studies fall short of being true computer simulations. It is appropriate to describe a few of these studies as they provide a good introduction into the later discussion of computer simulations.

Jose, Anderson, and Haight (1987) reported results from the StayWell program developed and marketed by Control Data Corporation. They showed that employees who smoke more than two packs a day, who are sedentary, who are 30% or more overweight, and who do not use seat belts have higher health care costs and absenteeism than employees without these risk factors. Combining this information with information about change in behavioral risk over a 6-year period, they estimated an annual savings of approximately $1.8 million to Control Data. The authors also conducted an analysis that looked at the expected savings from reducing these four risk factors that might be realized until the employee leaves the company or reaches the age of 65. They calculated, for example, that discounted to 1986 dollars, the net present value in cumulative savings was over $8,000 for a male at age 40 who reduced his behavioral risk.

Bertera (1991) analyzed data from the Du Pont Company. He was able to show a statistically reliable relationship between six behavioral risk factors (including smoking, elevated cholesterol, and high blood pressure) and absenteeism due to illness. He then estimated the cost of illness by estimating employee compensation, health care claims, and non–health care benefits. Based on the total Du Pont U.S. work force of 96,000 people, Bertera estimated that the total annual illness costs attributable to behavioral risks was $70,794,697, or $737 per person.

Finally, Golaszewski et al. (1992) analyzed data from the Travelers Insurance Company. They documented a relationship between chronological age and health care costs and, using data from an HRA, documented that the health promotion program at Travelers reduced the "health age" of program participants. These results were then used to estimate savings in health care costs that might result from a reduction in health age. The authors also estimated savings in the areas of absenteeism, life insurance, and productivity. Program costs were also estimated and cost-benefit ratios were projected from 1986 to the year 2000. The overall cost-benefit ratio was estimated to be 1 to 3.4, meaning that for every dollar invested in the health promotion programs 3.4 dollars were returned. The authors also examined the cost-benefit ratios under different assumptions (e.g., excluding the estimate of productivity benefits) and found that the cost-benefit ratio varied from 1.4 to 14.0 depending on what variables they included in their analysis. This study is particularly interesting because once they estimated program costs and program savings, they examined the impact of changes in some of their assumptions. As will be discussed below, this is called *sensitivity analysis* and is a powerful technique easily done in a computer simulation.

Computer simulations, however, are more than deterministic projections of future outcomes based on regression equations that use available data.

First, a computer simulation attempts to identify most if not all of the important variables that comprise the system or process under investigation. In doing so, computer simulations can include many more variables than typically reported in studies like those described above.

Second, as a computer simulation model is developed and used, it highlights assumptions that underlie the model. The developer or user is forced to recognize and defend the logic behind the model. If the model fails to pass a critical review by experts in the field or if it includes assumptions that cannot be adequately defended, then the model should be changed or the results will always be subject to question.

Finally, computer simulations allow you to easily ask "What if?" questions that are often used as aids to decision making and policy planning. This is a very attractive feature of computer simulations for several reasons. By systematically varying the values of independent and control variables it is possible to project outcomes under different scenarios. For example, one could develop "Base Case," "Worst Case," and "Best Case" scenarios to examine the range of values outcomes will take under different but realistic situations. If the differences are not large, there is less risk than if the differences are large. Or, if the results obtained for the worst case scenario are still acceptable, then the project can proceed with reasonable certainty that it will produce effects above the minimum threshold value. Under certain situations where stochastic, or Monte Carlo, procedures are used it is even possible to apply inferential statistics to outcome distributions associated with different scenarios to see if the projected differences are statistically significant.

Sensitivity analysis is a type of scenario analysis except that it examines the impact of changes in *one* variable at a time rather than in a set of variables. Golaszewski et al. (1992) did a sensitivity analysis when they calculated the cost-benefit ratio three ways, once with a 4% increase in employee productivity, once with a 25% increase, and once with a 0% increase. The cost-benefit ratios were 4.0, 14.0, and 1.4, respectively. This shows that cost-benefit ratios would appear to be highly sensitive to different estimates of productivity gains that might result from participation in a worksite-based health promotion program. Consequently, the accuracy of these estimates becomes critical for interpretation of results.

Two other decision-making aids that lend themselves to computer simulations and are relevant for the evaluation of health promotion programs at the worksite are *break-even analysis* and *capital budgeting*. These techniques can be used to estimate the economic effects of a health promotion program *prior* to implementation. The importance of these techniques should not be underestimated (Everly, Smith, & Haight, 1987).

Health promotion practitioners and researchers traditionally use a behavioral model to evaluate the impact of health promotion programs at the worksite. They are concerned with finding out if the program works by looking at such things as participation rates and changes in risk factors. But managers and corporate decision makers, who often must approve the substantial expenditures to start and maintain a health promotion program, most likely would use a financial model. They not only want to know if the program works—they also want to know how much the program costs and what the economic benefits will be.

Behavioral models are ex post because they evaluate program effects *after* the program has been implemented. Financial models are ex ante because they estimate the projected economic value of the program *prior* to a commitment of resources. Financial analysis is necessary if decision makers are to choose among alternative uses of financial resources. For example, managers cannot apply an ex post evaluation to the problem of deciding whether to increase sales and profits by buying a competitor or by building a new factory. They need information about the projected outcomes because it makes little sense to do both and then evaluate which approach was more profitable.

Break-even analysis focuses on the values certain variables must take to satisfy constraints placed on the model such that benefits derived from the program equal or exceed program costs (e.g., a cost-benefit ratio equal to or greater than 1.00). A senior manager, for example, may want to know what is the minimum percentage of employees who must participate in an exercise program for that program to break even. In this case, the simulation model solves for the employee participation rate for an exercise program with defined fixed and variable costs and fixed and variable benefits. Suppose the analysis suggests that for the program under consideration a minimum of 40% of eligible employees must participate in the exercise program if the program is to have a cost-benefit ratio of at least 1.00. Should the manager approve the expenditure of funds to launch the program? Because we know from experience that long-term participation rates in exercise programs rarely exceed 20% (Shephard, 1992), a projected participation rate in excess of 20% suggests that the program under consideration, which requires at least a 40% participation rate to break even, will probably cost more than it returns in dollar benefits. In this case the company would have to experience a higher than average participation rate to simply cover the costs of the program. Investing in this particular exercise program in this company, consequently, would require tolerance for a high risk of return on the investment. However, suppose the break-even analysis suggests that the program can break even if as few as 10% of employees participate. In this case, the manager can approve funding for the program with the expectation that participation rates equal to or greater than 10% would most likely be cost beneficial and yield a cost-benefit ratio greater than 1.00. That is, because the projected break-even point for employee participation should be easily attained, the risk attached to investment in this exercise program is low. By varying the values of input variables in a computer simulation, it is possible to examine the conditions where the health promotion program "breaks even." Decision makers then must decide whether or not those conditions are acceptable.

Besides break-even analysis, other capital budgeting techniques can easily be incorporated into computer simulations. Some of the more common techniques are the pay-back period, return on investment, internal rate of return, and net present value. The pay-back period is the amount of time in years required for a project to recover the initial investment. Return on investment is the ratio of annual cash flow derived from the project divided by project cost and multiplied by 100 to equal a percentage. Internal rate of return is the discount rate at which the project cost and expected cash flow produce a zero net present value. Net present value is a project's net contribution to the accumulation of cash, or the present value of the project's cash flow minus the project's initial cost. Net present value is generally recommended as the best technique (Brealey & Myers, 1991).

The material presented thus far has introduced the topic of computer simulations, showed why computer simulation is a promising approach to the evaluation of the economic impact of health promotion programs at the worksite, and reviewed studies that employed some but not all of the features of a computer simulation. The next section reports the results of two studies that used computer simulations and introduces original results from a new study.

Results of Computer Simulations of Worksite Health Promotion Programs

A comprehensive computerized search of the behavioral science and public health literature yielded only one paper (Patton, 1991) that reported the use of a computer simulation to evaluate the economic impact of health promotion programs at the worksite. A second paper, which is unpublished but widely disseminated (Terborg, 1988) and which has been presented at a national meeting (Terborg, 1990) will also be reviewed. Finally, the results of a Monte Carlo simulation that improves the Terborg (1988) model and incorporates more recent information will be presented (Terborg, 1993).

Patton's Model

Patton (1991) reports results from a computer simulation model that estimates the benefits and costs to a hypothetical organization of a 7-year-long health promotion program under a variety of assumptions regarding

Table 11.1
Model Variables With Base Case Values and the Range Used for Sensitivity Analysis: Patton (1991)

Variable	Base	Range
Annual participation rate	30%	10-60%
Annual turnover rate	15%	0-60%
Annual absenteeism costs/employee	$1,000	$100-$5,000
Annual health care costs/employee	$1,000	$500-$3,500
Retiree health care costs	0	0-$3,500
Turnover costs/event	$1,000	$100-$10,000
Cost of employee death	$1,500	$100-$10,000
Real discount rate	5%	0-10%
Real medical cost inflation	6%	0-10%

employee demographics and program effectiveness. Basic variables included in the model along with their initial values and the range of values examined in scenario and sensitivity analyses are presented in Table 11.1. Estimates of initial values and ranges were based on a review of the published literature, anecdotal communications by the author, and from an informal survey of 30 companies.

Benefits of the program are derived from changes in health care costs, turnover, absenteeism, and mortality. Employee turnover is assumed to decline by 2% in the first 2 years of an employee's exposure to the program, to decline 6% in years 3 to 6, and to terminate at a 10% reduction after 7 or more years of exposure. Absenteeism is assumed to decline by 5% in the first 2 years, 10% in years 3 to 6, and to terminate at a 15% reduction after 7 or more years of exposure. Mortality is assumed to decline by 1% during the first 2 years, 2% during years 3 to 6, and to terminate at a 3% reduction after 7 or more years of exposure. Health care costs are assumed to decline 2% during the first 2 years, 5% during years 3 to 6, and to terminate at a 10% reduction after 7 or more years of exposure to the program. The model assumes no costs and no benefits to nonparticipating employees. Fifty percent of the employees are male, gender is evenly distributed across age groups, and men and women are equally likely to participate in the program.

Patton conducted scenario analyses and sensitivity analyses (all values are in 1990 dollars). The base scenario (see Table 11.1) yielded a break-even point of $193. This means that if program costs are less than $193 per participant, the program will have a cost-benefit ratio greater than

1.00 and if program costs are more than $193 per participant the program will have a cost-benefit ratio less than one. Patton also examined the break-even point for four other scenarios. One scenario had a very young work force with high turnover. This scenario had a break-even point of $136. A second scenario had an older work force with low turnover. This yielded a break-even point of $242. A third scenario was created to model low program effectiveness and yielded a break-even point of $47. Finally, a high-effectiveness scenario yielded a break-even point of $328. As reported in the literature, program costs might be expected to range from $150 to $500 per participant (Erfurt, Foote, & Heirich, 1992; Shephard, 1992), or, assuming a 25% participation rate, between $37.50 and $125 per employee. Patton's results suggest that although health promotion programs would appear to be good investments, they might not break even in companies with a young work force and high turnover or in companies with programs that are only minimally successful in producing healthful changes in employee lifestyles.

Patton also conducted sensitivity analyses. A detailed description of his results will not be provided here, so the original paper should be studied for a complete understanding of Patton's findings. Within each of the five scenarios, Patton varied the values for participation rate, turnover rate, absenteeism cost, health care cost, retiree health care cost, turnover cost, cost of employee mortality, the discount rate, and medical cost inflation. Within the range of values used in the analysis, Patton concluded that differences in retiree health care costs, cost of mortality, medical cost inflation, and the discount rate had little impact on the break-even value. Differences in health care costs and participation rates had an intermediate impact. Differences in turnover rates, cost of turnover, and cost of absenteeism had the greatest impact on the break-even value.

Patton's computer simulation suggests that the characteristics of the company and its employees determine whether or not a health promotion program will be a good financial investment. Companies with low turnover, older employees who are costly to replace, and generous medical benefits are most likely to benefit. The results also indicate that between 50% and 75% of the economic benefits from a health promotion program come from productivity savings and not from health care savings.

Terborg's 1988 Model

Terborg (1988) conducted a computer simulation to evaluate the economic consequences of the Adolph Coors Wellness Program. The variables in his model and their values are presented in Table 11.2. Three scenarios were examined to simulate a base case, a worst case, and a best case set of assumptions. The actual values assigned to input variables were derived from data collected by Adolph Coors personnel and from the literature.

Table 11.2
Model Variables and Values Used for Base Case, Worst Case, and Best Case Scenarios: Terborg (1988)

Variable	Base case	Worst case	Best case
Number of salaried employees	1,952	1,952	1,952
Number of hourly employees	4,228	4,228	4,228
Average compensation: salaried	$57,047	$57,047	$57,047
Average compensation: hourly	$36,400	$36,400	$36,400
Productivity multiplier	3.50	2.00	5.00
Sick leave days: salaried	2.25	2.25	2.25
Sick leave days: hourly	4.10	4.10	4.10
Turnover rate: salaried	4.40%	4.40%	4.40%
Turnover rate: hourly	3.30%	3.30%	3.30%
Turnover cost/event: salaried	$8,173	$8,173	$8,173
Turnover cost/event: hourly	$2,506	$2,506	$2,506
Annual medical costs/employee	$1,000	$1,000	$1,000
Percent with health risk: salaried	41.50%	24.00%	59.00%
Percent with health risk: hourly	41.50%	24.00%	59.00%
Health risk medical cost differential	163.00%	133.00%	193.00%
Health risk sick leave differential	163.00%	133.00%	193.00%
Health risk productivity differential	98.25%	98.25%	98.25%
Inflation rate	3.50%	3.50%	3.50%
Discount rate	10.00%	10.00%	10.00%
Health risk participation rate	40.00%	29.00%	50.00%
No health risk participation rate	46.25%	16.50%	76.00%
Behavior change rate	36.00%	36.00%	36.00%
Health risk improvement: % no risk	100%	100%	100%
No risk improvement: medical	14.50%	10.00%	19.00%
No risk improvement: sick leave	14.50%	10.00%	19.00%
Reduction in turnover: salaried	20.00%	20.00%	20.00%
Reduction in turnover: hourly	10.00%	10.00%	10.00%
Full health cost impact in years	5.0	5.0	5.0
Full sick leave impact in years	2.0	2.0	2.0
Full productivity impact in years	2.0	2.0	2.0
Full turnover impact in years	1.0	1.0	1.0
Start-up costs	$95,200	$95,200	$95,200
Annual operating costs	$508,490	$227,184	$699,287
Corporate tax rate	37.89%	37.89%	37.89%

Terborg's model contains more variables than the model developed by Patton and incorporates developments from the field of personnel and human resource management to estimate productivity costs and benefits. Estimating the dollar value of changes in employee productivity, sick-leave, and turnover has been a major problem for health promotion researchers and practitioners. A promising solution is application of techniques developed by personnel and human resource management researchers to evaluate the economic impact of personnel programs dealing with such things as employee selection, training, and motivation. Cascio (1991) provides an excellent summary of this literature. Other contributions have been made by Schmidt, Hunter, and Pearlman (1982), Hunter and Schmidt (1983), and Boudreau (1983).

A detailed review of this literature is beyond the scope of this chapter, and some of the techniques and applications are still being debated (Hunter, Schmidt, & Coggin, 1988; Cronshaw & Alexander, 1991). Nevertheless, there is general agreement among researchers in this area that the dollar value of differences in employee behavior are substantially larger than originally thought. Hunter and Schmidt (1983) believe that the dollar value of employee productivity can be conservatively estimated as twice the employee's salary. Recent work by Judiesch, Schmidt, and Mount (1992) supports this position. Consequently, even small changes in employee on-the-job performance or in absenteeism due to illness and injury can translate into substantial dollar savings. Similarly, turnover costs have been estimated to be between 50% and 150% of annual salary for many managerial, technical, professional, clerical, and sales positions (Cascio, 1991). These findings lend support to Patton's results showing the importance of productivity in the evaluation of health promotion program effects.

Other features about Terborg's (1988) model that should be mentioned are the differentiation between costs and benefits for salaried versus hourly employees and for participants with excessive health risks versus those with no or minimal health risks, detailed analysis of both fixed and variable program costs, and recognition that only a fraction of program participants will actually be successful in making long-term behavior change. These features better reflect reality and should make results more valid. Terborg's model, however, does not consider retirement costs or employee age levels, which were included in Patton's (1991) model.

Terborg (1988) focused on four outcomes: productivity, absenteeism due to illness, voluntary turnover, and medical costs. The computer model was designed to make estimates in four areas: incremental costs that could be attributable to avoidable behavioral risk factors, potential annual savings and costs of the health promotion program, the number of employees who would have to participate in the program to break even, and the cumulative net present value of the program over a 10-year period.

Incremental costs associated with excessive behavioral risk factors ranged from $1,342 per at-risk employee in the worst case scenario to $6,316 per at-risk employee in the best case scenario. The cost was $3,419 in the base case. These figures suggest that in the base case, for example, each employee who is at elevated risk because of an unhealthy lifestyle costs the company $3,419 in extra medical costs, extra absenteeism, and in lost productivity. Medical costs accounted for between 10% and 25% of total excess costs, illness absenteeism accounted for between 5% and 10% of total excess costs, and losses in productivity accounted for between 65% and 85% of total excess costs. Turnover costs were not estimated in this analysis because turnover was not assumed to be related to health risk status. Rather, it was assumed that the existence of a health promotion program would reduce voluntary turnover because of its attractiveness as a benefit of employment. These results suggest that the hidden costs of employing people with unhealthy lifestyles can be substantial. When extrapolated to the entire Coors work force, potential incremental costs were $1,900,791, $282,936, and $5,874,180 in the base case, worst case, and best case scenarios, respectively.

The break-even analysis showed that the program would have a cost-benefit ratio greater than 1.00 if at least 4% of the total Coors work force would regularly participate in the wellness program's health promotion activities. Participation in the worst case scenario would have to exceed 9% of the work force to break even, and in the best case scenario, 2%. Actual participation by Coors employees was estimated at greater than 16% and as high as 76% in some activities. These results suggest that the Coors Wellness Program had a cost-benefit ratio of 3.75 in the base case, 1.24 in the worst case, and 8.33 in the best case scenario.

Net present value analysis was conducted to examine the net cash flow of savings (gross savings minus program costs) under different conditions. This analysis is important because the dollar gains from improved health often accrue in the future, whereas program costs, especially start-up costs, are high in the beginning. When looking at net present value analyses, it is useful to see how long it takes before the value of the cumulative net cash flow takes a positive value and what the total value is at certain points in time. Programs with very high early start-up costs and operating costs and low economic benefits might take several years before the net present value turns positive, if at all.

Results of the computer simulation showed that in the base case scenario, there was a net loss of $466,284 after the first year, but by the second year there was a net benefit of $259,255. After 5 years the total benefit to Coors was estimated to be $2,756,806. In the worst case scenario, the net present value never reached a positive number, with a net loss of $746,669 in the first year. Although the cash flow turned positive after a few years, indicating that on an annual basis the benefits of the program exceeded the annual operating costs, the gains were not large enough to overcome the substantial initial investment in start-up costs. Results from the best case scenario showed a positive value of $514,023 in the very first year and a total gain of over $12 million after 5 years.

Overall, the results showed that the Coors Wellness Program was a good investment of financial resources under most conditions. The report was positively received by senior managers, who initially supported the program in the absence of data and were anxious to learn if their commitment was financially sound. These results, however, are specific to Coors and, as Patton (1991) warned, results from one company might not generalize to other companies.

Terborg's 1993 Model

The values assigned to the variables used in Patton (1991) and Terborg (1988) were estimates of the mean, or average, values. Both assumed, for example, that the average annual health care cost per employee was $1,000. Use of the mean value is appropriate because in the long run, the best estimate of an aggregation of scores is the mean. Lynch, Teitelbaum, and Main (1992), however, pointed out that the distribution of health care costs is highly skewed and that the most expensive employee might have costs 100 to 500 times as much as the typical employee. Reporting only the mean cost obscures the fact that most employees would have costs below the mean and that only a few employees would have extreme costs above the mean.

Building on Lynch, Teitelbaum, and Main (1992), it follows that a computer simulation model that incorporated not only the mean value but some information about the range of values, the shape of the distribution, and the standard deviation would be a more realistic, and therefore a more accurate, evaluation tool.

Research on employee absenteeism (Hammer & Landau, 1981) and work performance (Hunter, Schmidt, & Judiesch, 1990) similarly shows that these values are not normally distributed but highly positively skewed. Hammer and Landau (1981), for example, found that in a sample of 112 employees, the number of involuntary days lost from work because of illness, injury, and other excusable reasons had a mean of 14.32 but ranged from 0 to 143 days. The median value was 5.71 and the mode was 2.00. The distribution of employee performance is positively skewed because although performance in theory would be normally distributed, in practice people who cannot perform at some minimal level are either not hired or are reassigned or terminated (Hunter, Schmidt, & Judiesch, 1990).

Advances in computer programs that do simulations made it possible to modify easily the 1988 Terborg model so that distributional assumptions could be input into the model and a technique known as "Monte Carlo" simulation could be used. In a Monte Carlo simulation, the computer makes multiple runs, or passes, through the input variables and generates a range of results that might occur in a particular situation. This feature is especially useful in the evaluation of health promotion programs at the worksite because of the skewed shape and high variability in values of key outcome variables. In Terborg's original model (1988), once the values were entered into the simulation, every time the model was run it would produce the exact same result. The only way to produce a different result was to change the values of the input variables through scenario analysis or sensitivity analysis. In Terborg's revised model (1993), once the mean *and* distribution of values are input, multiple runs of the program can produce different results. The variability in results stems from the variability inherent in the input variables. Consequently, with Monte Carlo analysis, the program is set to run a specified number of times and the output is a distribution of results. You are able to forecast the entire range of results and the likelihood of any specific result actually taking place.

The variables in Terborg's revised model (1993) and their mean values are presented in Table 11.3. Those variables that had distribution assumptions input into the model are presented in Table 11.4. The model was developed to simulate a hypothetical manufacturing company with 1,000 employees.

Besides inclusion of distributional information, the 1993 Terborg model contains a few additional changes. As shown in Table 11.3, the model now includes a variable that weights the percentage of change in productivity that can be attributed to a change in employee lifestyle and health risk status. In most cases, it is not reasonable to assign all of that gain to the employee. This change makes the 1993 model more conservative than the 1988 model. The model also distinguishes between medical cost inflation and wage inflation. Medical costs historically have risen faster than wages. This adjustment provides for a more accurate forecast. The estimate of program success has been changed to more accurately reflect change that can be attributed to the worksite health promotion program versus historical change. For example, we know that perhaps 3% to 5% of smokers will quit smoking on their own. A quit rate of 20% in a worksite smoking cessation program, then, might really reflect a net change of 15% to 17%. Finally, the values for many of the variables have been updated to reflect recent findings. The work of Lynch, Teitelbaum, and Main (1992), Wood, Olmstead, and Craig (1989), Erfurt, Foote, and Heirich (1991, 1992), Yen, Edington, and Witting (1992), Bertera (1990, 1991), Shephard (1992), and Golaszewski et al. (1992) guided the selection of values input into the model. Lynch (personal communication, April 1993) also made available information from two large companies with regard to absenteeism and health care costs. Specifically, for key variables, the values reported in the literature just cited were ranked from low to high and the median value was taken as the best estimate for the base case scenario.

Table 11.3
Model Variables and Mean Values Used for Base Case and High-Effectiveness Scenarios: Terborg (1993)

Variable	Base case scenario	High-effectiveness scenario
Number of salaried employees	250	250
Number of hourly employees	750	750
Average compensation: salaried	$55,000	$55,000
Average compensation: hourly	$32,000	$32,000
Productivity multiplier	2.00	2.50
Employee contribution: productivity gains	75.00%	94.00%
Sick leave days: salaried	3.00	3.75
Sick leave days: hourly	5.00	6.25
Turnover rate: salaried	5.00%	6.25%
Turnover rate: hourly	16.00%	20.00%
Turnover cost/event: salaried	$27,500	$34,375
Turnover cost/event: hourly	$2,667	$3,334
Annual medical costs/employee	$1,500	$1,875
Percent with health risk: salaried	30.00%	37.50%
Percent with health risk: hourly	40.00%	50.00%
Health risk medical cost differential	150.00%	187.50%
Health risk sick leave differential	135.00%	168.75%
Health risk productivity differential	96.00%	95.00%
Medical cost inflation rate	12.00%	12.00%
Wage inflation rate	3.00%	3.00%
Nominal discount rate	10.00%	10.00%
Health risk participation rate: salaried	25.00%	31.25%
Health risk participation rate: hourly	15.00%	18.75%
No health risk participation rate: salaried	30.00%	37.50%
No health risk participation rate: hourly	20.00%	25.00%
Program success rate (net)	20.00%	25.00%
Health risk improvement: % no risk	100%	100%
No risk improvement: medical	5.00%	6.25%
No risk improvement: sick leave	5.00%	6.25%

(continued)

Table 11.3 *(continued)*

Variable	Base case scenario	High-effectiveness scenario
Reduction in turnover: salaried	10.00%	12.50%
Reduction in turnover: hourly	5.00%	6.25%
Full health cost impact in years	4.0	4.0
Full sick leave impact in years	2.0	2.0
Full productivity impact in years	2.0	2.0
Full turnover impact in years	2.0	2.0
Start-up costs	$100,000	$100,000
Annual operating costs/employee	$100	$100
Corporate tax rate	34.00%	34.00%

Table 11.4
Standard Deviations and Ranges
for Distributional Assumptions: Terborg (1993)

Variable	Standard Deviation	Range
Productivity multiplier: salaried	0.7	0.9-4.54
Productivity multiplier: hourly	0.7	0.9-3.36
Employee contribution: productivity	7.5%	0-100%
Sick leave days: salaried	6.4	0-146
Sick leave days: hourly	6.4	0-146
Turnover costs/event: salaried	$2,750	$4,500k-$110,000
Turnover costs/event: hourly	$267	0-$16,000
Annual medical costs/employee	$3,000	0-$50,000

In addition to the base case, a high-effectiveness scenario was created. With the exception of number of employees, salary, inflation, and discount rates, the at-risk improvement factor, time in years for the health promotion program to have full impact, and operational cost variables, all other variables were adjusted by 25% in a direction that would make the program more effective. This adjustment was well within reasonable limits of values reported in the literature and presents a realistic simulation of a highly effective health promotion program at the worksite.

Summary results for 500 runs of the base case and high-effectiveness case scenarios are presented in Table 11.5. Outcome variables were medical cost savings, productivity savings, sick leave savings, turnover savings, savings per participant, the cost-benefit ratio, and the break-even participation rate. As seen in Table 11.5, for both scenarios two patterns exist. First, the dollar value of savings in employee productivity is substantially larger than savings found in other categories, replicating Terborg (1988) and Patton (1991). And, second, for most of the outcome variables the observed values show considerable variability and a tendency toward a positive skew. Savings per participant, for example, in the base case has a mean savings of $313 but a range from $187 to $3,003 and a standard deviation of $73.

Although some variability would be expected given the distributional assumptions placed in the model, the magnitude of variability is still surprising. If each run of the simulation model estimates what could happen in any given year, then the whole logic behind doing an evaluation

Table 11.5
Results of Monte Carlo Simulation
for Base Case and High-Effectiveness Scenarios: Terborg (1993)[1]

	Mean	Standard Deviation	Range
Base case scenario:			
Medical savings	$10,281	$15,966	$119-$195,358
Productivity savings	$39,766	$9,495	$17,027-$71,872
Sick leave savings	$3,131	$2,960	$224-$22,908
Turnover savings	$12,701	$1,021	$10,008-$17,004
Total savings	$65,879	$18,915	$32,122-$245,300
Savings per participant	$313	$73	$187-$3,003
Cost-benefit ratio	0.55	0.10	0.34-1.85
Break-even participation rate	38%	7%	11%-62%
High-effectiveness scenario:			
Medical savings	$33,139	$52,856	$441-$632,523
Productivity savings	$128,040	$25,062	$64,051-$202,105
Sick leave savings	$11,562	$8,701	$1,576-$60,263
Turnover savings	$29,359	$2,475	$21,365-$36,211
Total savings	$202,100	$58,656	$112,653-$805,611
Savings per participant	$770	$232	$425-$3,177
Cost-benefit ratio	1.69	0.51	0.93-6.97
Break-even participation rate	16%	3%	4%-27%

[1]Results based on 500 runs. Dollars indicate estimates of total annual savings for a hypothetical manufacturing company with 1,000 employees.

of a worksite health promotion program needs to be rethought, especially as it relates to the base case scenario.

Consider the task of the director of a corporate health promotion program similar to that depicted in the base case scenario who has been instructed to conduct a 1-year long evaluation of the program for review by senior management. According to the computer simulation, the estimate is that the program will produce a savings of $313 per participant before subtracting out any costs of the program. But it could happen that the calculated figure for that year turns out to be as low as $200 or as high as $400. Suppose $313 with a cost-benefit ratio of 0.55 is acceptable to senior management with a proposal to make up some of the unrecovered program cost with a user fee of $100 per year. But, a forecasted savings of $200 is unacceptable and participants are unwilling to pay more than $100 per year for the program. If, because of the inherent variability in results that might be observed from one year to the next, the director is unlucky enough to find a savings of $200 per participant after completion of an expensive and demanding program review, senior management might decide, incorrectly, to drastically reduce or even eliminate the program. On the other hand, if a savings of $400 per participant is calculated, senior management might decide to keep the program as it is and actually cost the company money when it thinks it is close to breaking even.

Results obtained in the high-effectiveness scenario are less ambiguous to the decision maker even though the same magnitude of variability is present. This is because only in rare instances would a review of the program produce such a low estimate of benefits that continuation of the program would be seriously debated.

The point of this discussion is that if program effects are as variable as the computer simulation suggests they are, then attempts to actually conduct a program evaluation for the purposes of documenting economic benefits may be unjustified in most cases. This does not imply that research and documentation should stop. It does imply, however, that the results of any one evaluation of any one worksite health promotion program must be interpreted with considerable caution and care.

Another pattern that emerges from the results is the magnitude of differences between the base scenario and the high-effectiveness scenario. Given the costs of the program at $100,000 per year and start-up costs of $100,000, the results in the base case are not as encouraging as those reported by Patton (1991) or Terborg (1988). In fact, it would be difficult to justify on economic grounds continued support of the program. The mean cumulative net present value for this program after 5 years was computed to be a negative $235,627. However, the high-effectiveness scenario, though still not as positive as earlier reports in the literature (Shephard, 1992), is quite respectable with a mean cost-benefit ratio of 1.69, a break-even participation rate of 16%, and a 5-year cumulative net present value of $111,212.

A detailed sensitivity analysis has not yet been conducted on the 1993 Terborg model. However, preliminary analyses suggest that differences in medical cost inflation, the discount rate, and the corporate tax rate have relatively little effect, with the latter being more important than the other two. The impact of employee participation rates, program success rates in producing long-term reductions in health risks, and use of employee fees seem to have a greater impact and represent areas more under the control of health promotion managers. Finally, as shown by Patton (1991), potential economic benefits vary substantially as a function of organizational characteristics, employee health status, program effectiveness, and level of medical benefits.

Conclusions and Directions for Future Research

Computer simulation is a promising technique for the evaluation of the economic impact of health promotion programs at the worksite. This technology represents a viable alternative to costly and difficult program evaluations that require the use of experimental designs and longitudinal data collection. It can be used ex post to estimate cost-benefits of an existing program, as was done with the Coors Wellness Program (Terborg, 1988). It also can be used ex ante to evaluate a program that is in the planning stage.

The three simulations reviewed in this chapter suggest that health promotion programs may be cost-beneficial, that economic benefits primarily come from increases in productivity and less so from decreases in health care costs, and that program effects can vary considerably both within an organization from year to year and between different organizations.

The application of computer simulations to the evaluation of worksite health promotion programs has begun only recently, and there is considerably more work to be done. The results should be interpreted with some caution as none of the models have been independently validated. However, the results are well within the range of existing empirical research, and Patton (1991) and Terborg (1988, 1993) found similar patterns using independent computer simulation models.

In closing, the following questions are offered as research priorities to guide future work in this area:

1. What are the assumptions that need to be made prior to the design of a computer simulation?
2. What are the most important variables that need to be considered as inputs, control variables, and outputs?
3. What algorithms or rules should be specified for linking input and control variables to outputs?

4. What range of values do we assign to input and control variables? How do we know these estimates are valid?
5. Can we validate the simulation by demonstrating correspondence between outcomes empirically derived and outcomes predicted by the model?
6. Are results of computer simulations useful as decision aids to executives and policy planners who make decisions regarding health promotion programs at the worksite?
7. Do computer simulations help researchers identify important research questions that are amenable to subsequent empirical verification?
8. Do computer simulations provide for a realistic understanding of the structure and internal processes of health promotion programs at the worksite?

Acknowledgments

David Dusseau, PhD, and Robert Clemen, PhD, provided assistance regarding computer simulation programs. Financial support for the preparation of this chapter was provided through a gift from Carolyn S. Chambers to the College of Business Administration at the University of Oregon and Grant #HL45548 from the National Heart, Lung and Blood Institute.

References

Bernacki, E.J., & Baun, W.B. (1984). The relationship of job performance to exercise adherence in a corporate fitness program. *Journal of Occupational Medicine*, **26**, 529-531.

Bertera, R.L. (1990). The effects of workplace health promotion on absenteeism and employment costs in a large industrial population. *American Journal of Public Health*, **80**, 1101-1105.

Bertera, R.L. (1991). The effects of behavioral risks on absenteeism and health-care costs in the workplace. *Journal of Occupational Medicine*, **33**, 1119-1124.

Blair, S.N., Piserchia, P.V., Wilbur, C.S., & Crowder, J.H. (1986). A public health intervention model for worksite health promotion. *Journal of the American Medical Association*, **255**, 921-926.

Boudreau, J.W. (1983). Economic considerations in estimating the utility of human resource productivity improvement programs. *Personnel Psychology*, **36**, 551-576.

Brealey, R.A., & Myers, S.C. (1991). *Principles of corporate finance*. New York: McGraw-Hill.

Cascio, W.F. (1991). *Costing human resources: The financial impact of behavior in organizations*. Boston: PWS-Kent.

Cronshaw, S.F., & Alexander, R.A. (1991). Why capital budgeting techniques are suited for assessing the utility of personnel programs: A reply to Hunter, Schmidt, and Coggin (1988). *Journal of Applied Psychology*, **76**, 454-457.

Erfurt, J.C., Foote, A., & Heirich, M.A. (1991). The cost-effectiveness of work-site wellness programs for hypertension control, weight loss, and smoking cessation. *Journal of Occupational Medicine*, **33**, 962-970.

Erfurt, J.C., Foote, A., & Heirich, M.A. (1992). The cost-effectiveness of worksite wellness programs for hypertension control, weight loss, smoking cessation, and exercise. *Personnel Psychology*, **45**, 5-27.

Everly, G.S., Smith, K.J., & Haight, G.T. (1987). Evaluating health promotion programs in the workplace: Behavioral models versus financial models. *Health Education Research*, **2**, 61-67.

Fielding, J.E. (1984). Health promotion and disease prevention at the worksite. *Annual Review of Public Health*, **5**, 237-265.

Glasgow, R.E., & Terborg, J.R. (1988). Occupational health promotion programs to reduce cardiovascular risk. *Journal of Consulting and Clinical Psychology*, **56**, 365-373.

Golaszewski, T., Snow, D., Lynch, W., Yen, L., & Solomita, D. (1992). A benefit-to-cost analysis of a worksite health promotion program. *Journal of Occupational Medicine*, **34**, 1164-1172.

Hammer, T.H., & Landau, J. (1981). Methodological issues in the use of absence data. *Journal of Applied Psychology*, **66**, 574-581.

Hatcher, M., & Rao, N. (1988). A simulation based decision support system for a health promotion center. *Journal of Medical Systems*, **12**, 11-29.

Hunter, J.E., & Schmidt, F.L. (1983). Quantifying the effects of psychological interventions on employee job performance and work-force productivity. *American Psychologist*, **38**, 473-478.

Hunter, J.E., Schmidt, F.L., & Coggin, T.D. (1988). Problems and pitfalls in using capital budgeting and financial accounting techniques in assessing the utility of personnel programs. *Journal of Applied Psychology*, **73**, 522-528.

Hunter, J.E., Schmidt, F.L., & Judiesch, M.K. (1990). Individual differences in output variability as a function of job complexity. *Journal of Applied Psychology*, **75**, 28-42.

Jose, W.S., Anderson, D.R., & Haight, S.A. (1987). The StayWell strategy for health care cost containment. In J.P. Opatz (Ed.), *Health promotion evaluation: Measuring the organizational impact* (pp. 15-34). Stevens Point, WI: National Wellness Institute.

Judiesch, M.K., Schmidt, F.L., & Mount, M.K. (1992). Estimates of the dollar value of employee output in utility analyses: An empirical test of two theories. *Journal of Applied Psychology*, **77**, 234-250.

Lewandowsky, S., & Murdock, B.B. (1989). Memory for serial order. *Psychological Review*, **96**, 25-57.

Lynch, W.D., Teitelbaum, H.S., & Main, D.S. (1992). Comparing medical costs by analyzing high-cost cases. *American Journal of Health Promotion*, **6**, 206-213.

Opatz, J.P. (Ed.) (1987). *Health promotion evaluation: Measuring the organizational impact*. Stevens Point, WI: National Wellness Institute.

Opatz, J.P., Chenoweth, D., & Kaman, R. (1991). *Economic impact of worksite health promotion programs*. Indianapolis: Association for Fitness in Business Publications.

Pallin, A., & Kittell, R.P. (1992). Mercy Hospital: Simulation techniques for ER processes. *Industrial Engineering*, **13**, 35-37.

Patton, J.P. (1991). Work-site health promotion: An economic model. *Journal of Occupational Medicine*, **33**, 868-873.

Schmidt, F.L., Hunter, J.E., & Pearlman, K. (1982). Assessing the economic impact of personnel programs on workforce productivity. *Personnel Psychology*, **35**, 333-347.

Stoll, R.J. (1983). Nations at the brink: A computer simulation of governmental behavior during serious disputes. *Simulations and Games*, **14**, 179-200.

Shephard, R.J. (1992). A critical analysis of work-site fitness programs and their postulated economic benefits. *Medicine and Science in Sports and Exercise*, **24**, 354-370.

Terborg, J.R. (1988). *Cost benefit analysis of the Adolph Coors wellness program*. Unpublished manuscript, University of Oregon, Eugene.

Terborg, J.R. (1990, May). *The application of behavioral costing techniques to the evaluation of health promotion programs at the worksite: A cost benefit analysis of the Adolph Coors wellness program*. Paper presented at the First Conference on the Economic Impact of Employee Health Promotion Programs, Association for Fitness in Business, Fort Worth, TX.

Terborg, J.R. (1993, May). *Computer simulations*. Paper presented at the Second Conference on the Economic Impact of Employee Health Promotion Programs, Association for Worksite Health Promotion, Buffalo, NY.

U.S. Department of Health and Human Services Public Health Service (1993). 1992 national survey of worksite health promotion activities: summary. *American Journal of Health Promotion*, **7**, 452-465.

Warner, K.E., Wickizer, T.M., Wolfe, R.A., Schildroth, J.E., & Samuelson, M.H. (1988). Economic implications of workplace health promotion programs: Review of the literature. *Journal of Occupational Medicine*, **30**, 106-112.

Whicker, M.L., & Sigelman, L. (1991). *Computer simulation applications: An introduction*. London: Sage.

Wood, E.A., Olmstead, G.W., & Craig, J.L. (1989). An evaluation of lifestyle risk factors and absenteeism after two years in a worksite health promotion program. *American Journal of Health Promotion*, **4**, 128-133.

Yen, L.T., Edington, D.W., & Witting, P. (1991). Associations between health risk appraisal scores and employee medical claims costs in a manufacturing company. *American Journal of Health Promotion*, **6**, 46-53.

Yen, L.T., Edington, D.W., & Witting, P. (1992). Prediction of prospective medical claims and absenteeism costs for 1284 hourly workers from a manufacturing company. *Journal of Occupational Medicine*, **34**, 428-435.

Appendix

Worksite Health Promotion in Health Care Reform

A Position Statement of the Worksite Health Promotion Alliance

Reducing health care costs without lowering quality is a national priority. Two of the most effective ways to reduce these costs are to encourage Americans to adopt better health habits and to educate them about enlightened use of health care services. The best results in producing meaningful changes in health habits have come from worksite health promotion programs. However, economic incentives are the key to getting and keeping employers involved. Health care reform should encourage employer initiatives in health promotion—and not inadvertently undermine them.

This document summarizes the positive role that worksite health promotion can play in achieving the goals of health care reform. It was developed by the **Worksite Health Promotion Alliance**, *a coalition of organizations and companies representing millions of American workers and their dependents.*

Effectiveness of the Worksite as a Setting for Health Promotion

Half—or more—of all disease is caused by poor health behaviors.[1] The power of health promotion is that it reduces health care costs by keeping people healthy. And the worksite has proven to be one of the most effective settings for promoting health for all Americans.

Most of the 110 million people who make up the American work force spend the major portion of their day at the worksite.[2] In addition, millions of their dependents are indirectly influenced by worksite programs. Work settings thus present an extraordinarily broad-based opportunity for reaching a huge number of Americans.

Up to 30% of all employer-paid health care costs are due to unhealthy lifestyle habits, which can be significantly improved by worksite health promotion programs. Health care reform could enhance—or destroy—proven employer successes in modifying these habits, with their commensurate reductions in cost.

A mechanism for health promotion already exists in many worksites, especially those with large employee groups. Alarmed by the rising cost of company-provided health insurance, American business has made significant progress in the last 20 years in educating employees about these issues and using the worksite as a place to promote healthy behaviors. Worksite health promotion programs have also taken the lead in providing education to employees for the responsible use of the health care system, resulting in decreased utilization and medical costs.

Worksite health promotion is a particularly effective way to reach people at high risk. Typically the people who need health promotion the most—those who are the least aware of their unhealthy lifestyles and the least motivated to change—will not go out of their way to get it. The worksite provides a convenient setting for educating them and for offering the peer support they need to start and maintain healthy habits. With its captive audience of a broad cross-section of people, the worksite provides a way to reach high-risk workers who are not likely to volunteer for community-based programs. Worksite-based programs, therefore, have a unique advantage.

Health promotion reinforces other business efforts aimed at improving worker productivity and effectiveness. Obvious examples include initiatives to reduce worker compensation claims, decrease risk of injury, improve safety, increase productivity, enhance quality, and encourage the team building process. Health promotion is also a natural ally of the company benefits department and the insurance provider—often joining them in programs that affect health care costs related to issues like prenatal care, quicker return to work from disability, and more appropriate utilization of the health care system.

The worksite presents a powerful way to positively impact the health culture of Americans. People trying to adopt healthy lifestyle habits are more likely to succeed if they have the support of their peers and friends. The worksite provides a natural, highly leveraged opportunity for this kind of positive reinforcement. Many companies believe that the most important contribution to improving health is the changing of organizational norms to support good health habits. Indeed, changes in the work culture may contribute strongly to the effectiveness of worksite health promotion programs.

Worksite health promotion has proven to be very effective in reducing both lifestyle-related health risk and the resulting demands on the medical care system.[3] In a comprehensive review of studies conducted on a wide range of worksite health promotion programs, leading researchers concluded that such programs reduced annual health care costs per employee by as much as $865.[4]

Strategies for Including Worksite Health Promotion in Health Care Reform

Over the past 20 years, employers, on their own initiative, have introduced worksite health promotion into their companies to reduce their costs for health care. These activities can be preserved and further enhanced by the inclusion of health promotion in health care reform. Health promotion activities by the employer at the worksite should be strongly encouraged. These activities include

- health risk assessments,
- smoking cessation programs,
- healthy pregnancy programs,
- blood pressure reduction programs for those suffering from hypertension,
- cholesterol reduction programs for those with high cholesterol,
- physical fitness programs,
- healthy eating programs, and
- screenings for disease conditions as recommended in *The Guide to Clinical Preventive Services*.[5]

Companies have also learned that the impact of worksite health promotion programs is enhanced by creating a total corporate culture that supports healthy behaviors. They have instituted nonsmoking policies, created heart-healthy menu selections in the company cafeteria, and have established management strategies designed to minimize company-induced stress.

Economic Incentives That Work

Historically, experienced-based rating has been a successful economic incentive for companies that promote employee health and well-being. The alliance recommends that regardless of the system enacted, health care reform must include economic incentives for employers to maintain or establish worksite health promotion and disease prevention programs.

Continued Tax Deductibility

It is essential that health promotion programs remain fully tax-deductible to the company without being considered as income to individual participants. Deductibility should apply to companies who provide worksite health promotion programs that go beyond the activities covered in the "core" health care plan. Taxing this benefit would severely curtail worksite health promotion efforts.

Discounts for Employers With Worksite Health Promotion Programs

For employers whose health care coverage is required to be provided by a health alliance, we recommend offering a three-tier system of ratings, based on the level of health promotion efforts. This rating system should be implemented as a basis for incentives for companies to offer these programs at the worksite:

Tier One	No discount per employee for companies *without* a qualified worksite health promotion program.
Tier Two	A discount per employee for companies with a *basic* qualified worksite health promotion program.
Tier Three	A greater discount per employee for companies with a *comprehensive* qualified worksite health promotion program.

Detailed standards for determining a qualified program are available from the Worksite Health Promotion Alliance.

Assistance and Training for Small Business

Well over half of Americans with jobs work for small companies. These companies usually offer a lower level of health promotion activities than large companies. Therefore, the proposed National Health Board or the Department of Health and Human Services should work with small business groups, private vendors, and worksite health promotion organizations to design and implement programs to promote healthy lifestyles among employees of small companies. This initiative should include financial incentives, education and awareness campaigns, guidance on how to start programs, and a clearinghouse of information on successful programs, including a resource list of agencies and vendors who provide worksite health promotion services, programs, and materials.

* * *

Worksite health promotion programs are widely distributed throughout America's business community. They produce results. Health care reform should ensure that this highly effective channel for promoting healthy lifestyles is not undermined. Health promotion at the worksite must be sustained—and indeed, strengthened—to ensure that health care reform succeeds.

June 1993
The Worksite Health Promotion Alliance
Chicago, Illinois
Revised 10/11/93

To add your support to the Worksite Health Promotion Alliance, contact one of the founding members:

Bill Whitmer, Wellness South, Inc. (205-988-4441)
Carson Beadle, The Health Project (212-345-7004)
Joseph Burns, Business & Health (201-358-7208)
Tracey Cox, Presbyterian Hospital of Dallas (214-345-4656)
Janet Edmunson, Association for Worksite Health Promotion (617-332-9600)
William Horton, Fitness Systems (310-312-2988)
Harold Kahler, Wellness Councils of America (402-572-3590)
Robert Kaman, University of North Texas Health Science Center at Fort Worth (817-735-2670)
Michael O'Donnell, American Journal of Health Promotion (313-650-9600)
Kenneth R. Pelletier, Stanford University School of Medicine (415-723-1000)
Ron Goetzel, Johnson & Johnson Health Management (202-965-6929)
Jeremy Rifkin, Foundation on Economic Trends (202-466-2823)
Dennis Richling, Union Pacific Railroad (402-271-4326)
Susan Seidler, Washington Business Group on Health (202-408-9320)
Neal Sofian, Group Health/Puget Sound (206-287-4396)

References

1. Fries, J.F., Koop, C.E., Beadle, C.E., Cooper, P.O., England, M.J., Greaves, R.F., Sokolov, J.J., Wright, D., & The Health Project Consortium. Reducing Health Care Costs By Reducing the Need and Demand for Medical Services, *New England Journal of Medicine,* Vol. 329, No. 5, 1993.
2. *Healthy People 2000: National Health Promotion and Disease Prevention Objectives,* Occupational Safety and Health, Section 10, p. 296.
3. Pelletier, K.R. A Review and Analysis of the Health and Cost-Effectiveness Outcome Studies of Comprehensive Health Promotion and Disease Prevention Programs at the Worksite: 1991-1993 Update, *American Journal of Health Promotion,* Vol. 8, No. 1, 1993.
4. Opatz, J., Chenoweth, D., & Kaman, R.L., *Economic Impact of Worksite Health Promotion,* Association for Worksite Health Promotion Publications, 1990, Northbrook, IL.
5. U.S. Preventive Service Task Force. *The Guide to Clinical Preventive Services,* Williams and Wilkins, 1988, Baltimore, Maryland.

Index

About the Editor

Robert L. Kaman, PhD, is a recognized authority in the field of health promotion, leading the effort to research the economic impact of worksite health promotion, chairing symposia, and contributing to the scholarly and professional literature.

Kaman has been an active member of the Association for Worksite Health Promotion since 1979. He served on the board of directors from 1987 through 1989 and in 1992 was elected president. Kaman is also a member of the American College of Sports Medicine and the American Osteopathic Association of Sports Medicine.

Kaman is an associate professor in the department of physiology and an adjunct associate professor in the department of public health/preventive medicine at the University of North Texas Health Science Center at Fort Worth. He earned his PhD in biochemistry from the Virginia Polytechnic Institute in 1969 and is working toward a law degree at Texas Wesleyan University.

About the Contributors

William B. Baun, MS, FAWHP, is the manager of the Tenneco Health Promotion Department and has been responsible for the establishment of over 20 health promotion/fitness programs at Tenneco in a variety of white and blue collar settings. He has numerous publications and serves as an adjunct professor at the University of Texas Public Health School and the University of Houston.

Steven N. Blair, PhD, is the Director of Epidemiology and Clinical Applications at the Cooper Institute for Aerobics Research. Blair is widely known for his research that defined the quantity of physical activity and its relative effect on cardiovascular health.

David Chenoweth, PhD, FAFB, is currently a professor and director of Worksite Health Promotion Studies at East Carolina University. He is also the president of Health Management Associates, where he provides health promotion and health care cost management strategies to businesses.

D.W. Edington, PhD, is the director of the Fitness Research Center at the University of Michigan. He is also a professor and director of the division of kinesiology. His research, in collaboration with Dr. Louis Yen, is focused on the relationships between health behaviors and risks, health utilization and costs, and productivity and quality of life.

Thomas J. Golaszewski, EdD, MS, is a nationally recognized leader and researcher in the health promotion field and has over 20 years of experience in schools, universities, corporations, and public health institutions. Dr. Golaszewski is vice president for research and product development for DINE Systems, Inc., an Amherst, New York, based health software development firm. He also serves as senior research associate for The Prevention Resource Center, S.U.N.Y. College at Buffalo.

Wendy D. Lynch, PhD, is principal of Health Decisions, Inc., and is a consultant for several corporations in the areas of claims analysis and cost containment. She has authored several papers and publications and is a frequent speaker on health care costs and the relationship between health and medical utilization.

Roy J. Shephard, MD, PhD, is presently professor of applied physiology in the School of Physical and Health Education and Department of Preventive Medicine and Biostatistics, Faculty of Medicine at the University of

Toronto. He holds an MD and PhD from London University, honorary doctorates in physical education from Gent University and the University of Montreal, and is a former president of the Canadian Association of Sports Sciences and the American College of Sports Medicine.

James R. Terborg, PhD, is the associate dean in the College of Business Administration and director of the Institute of Industrial Relations, both at the University of Oregon. He is currently coprincipal investigator on a 5-year grant from the National Heart, Lung and Blood Institute, looking at occupational health promotional programs.

R. William Whitmer, RPh, MBA, is founder, president, and CEO of Wellness South, Inc. He has designed and produced comprehensive employee/dependent health promotion programs for dozens of private and public sector clients.

MEMBERSHIP FACT SHEET

AWHP Membership Benefits

Individual members and the official representatives for company memberships receive these "tangible" benefits:

- A subscription to the AWHP's *Worksite Health*
- A subscription to the AWHP *Action* newsletter
- The annual Who's Who Membership Directory and Resource Guide (an exclusive benefit) and free inclusion if membership status is current as of February 1
- Automatic membership in regional and local sections
- Eligibility to subscribe to the JOB Opportunity Bureau
- Preferred rates for the AWHP conference and events

In addition, the Association will offer and arrange special programs or services for AWHP members only.

Associate member companies receive the following additional benefits:

- Eligibility to exhibit at the Annual Conference
- Preferred rate (an additional 10% reduction of the low member price) on mailing list rental

AWHP Membership Categories

The association has a category of membership available to meet your situation. Following is a listing of the types of memberships provided. Please note that only the Professional Member has the right to hold office or vote on Association matters.

1. **Individual Memberships.** These memberships are in the name of the individual rather than the company or organization.
   A. **Professional Member** is for individuals who derive income from a health promotion profession by providing educational development, management services, or evaluations of health promotion programs.
   B. **General Member** is for individuals who have an active interest in health promotion, but do not derive income from a health promotion profession.
   C. **Student Member** is for full-time undergraduate and graduate students enrolled in a program of study related to health promotion. (AWHP also provides a student chapter program at qualified institutions. Please call the AWHP office for more information.)
2. **Company Memberships.** These memberships are in the name of a firm, business or organization. The company/organization designates the person or persons who will represent them with the Association and pays annual dues

for each representative. Memberships may be transferred by the company/organization upon written notice to AWHP.

A. **Associate Member** is available to any firm, business, or corporation engaged in selling products or services to members of the Association.

B. **Company/Organization Member** is available to any firm, business, not-for-profit organization, or institution with an active interest in health promotion or that may be helpful in carrying out the objectives of the Association.

AWHP Application for Membership

(Please type, print, or attach business card)

Date of Application __________ Name ______________________ Nickname __________

(First) (M.I.) (Last)

Title ______________________ Company/Organization ______________________

Address ______________________ City __________ State ________ Zip ________

Phone (_____) ______________________ Fax (_____) ______________________

Please check the membership you are applying for and submit the appropriate annual fee in U.S. dollars or the equivalent. **Memberships are based on the calendar year.**

❑ Professional Member—$130

❑ Associate Member—$350

❑ General Member—$130

❑ Student Member*—$70

❑ Company/Organization Member—$250

*A student application must be accompanied by a letter from the registrar's office or a current transcript.

❑ Check enclosed for $ __________

❑ Charge $ ________ to my ❑ Visa ❑ MasterCard Acct. # ______________________

Exp. Date __________ Signature ______________________

Sponsor's name/who introduced you to AWHP ______________________

(Optional)

You will begin receiving services upon receipt of payment.
Please allow 4–6 weeks for initial receipt of publication.

If you have any questions regarding your membership services, please call AWHP at (708) 480-9574, fax (708) 480-9282.

Mail completed application to AWHP, 60 Revere Dr., Ste. 500, Northbrook, IL 60062.

More authoritative books for the worksite

Economic Impact of Worksite Health Promotion

A publication of the Association for Worksite Health Promotion

Joseph P. Opatz, PhD, Editor

1994 • Cloth • 272 pp • Item BOPA0436
ISBN 0-87322-436-1 • $30.00 ($40.50 Canadian)

Economic Impact of Worksite Health Promotion is an essential tool in helping you to evaluate and modify health promotion programs. In this book, business and health promotion experts address important theoretical considerations affecting worksite health promotion, review and evaluate the effectiveness of current worksite health promotion studies, lead you through the steps in planning a program evaluation, and examine specific program studies from diverse worksite settings.

Guidelines for Employee Health Promotion Programs

Association for Fitness in Business

Jean Storlie, MS, RD, William B. Baun, MS, and William L. Horton, MBA

1992 • Paper • 152 pp • Item BAFB0351
ISBN 0-87322-351-9 • $26.00 ($34.95 Canadian)

No other book takes you step by step through the start-up phases of an employee health promotion initiative. From the initial needs analysis to the program's mission statement, marketing, and year-end evaluation, you'll find the information you need to develop and implement a quality employee health promotion program.

Place your order using the appropriate telephone number and address shown in the front of this book, or **call toll-free in the U.S. 1-800-747-4457.**

Prices subject to change.

Human Kinetics

The Information Leader in Physical Activity

2335